ENVIRONMENTAL REMEDIATION TECHNOLOGIES, REGULATIONS AND SAFETY

FOREST CERTIFICATION AND SUSTAINABLE MANAGEMENT

PROGRAMS, STANDARDS AND TECHNIQUES

Environmental Remediation Technologies, Regulations and Safety

Additional books in this series can be found on Nova's website under the Series tab.

Additional E-books in this series can be found on Nova's website under the e-books tab.

Environmental Science, Engineering and Technology

Additional books in this series can be found on Nova's website under the Series tab.

Additional E-books in this series can be found on Nova's website under the e-books tab.

ENVIRONMENTAL REMEDIATION TECHNOLOGIES, REGULATIONS AND SAFETY

FOREST CERTIFICATION AND SUSTAINABLE MANAGEMENT

PROGRAMS, STANDARDS AND TECHNIQUES

JOHN G. ROBLES
AND
CHARLOTTE A. SAVAGE
EDITORS

Nova Science Publishers, Inc.
New York

For permission to use material from this book please contact us:
Telephone 631-231-7269; Fax 631-231-8175
Web Site: http://www.novapublishers.com

Additional color graphics may be available in the e-book version of this book.

Library of Congress Cataloging-in-Publication Data

ISBN: 978-1-62081-206-8

Published by Nova Science Publishers, Inc. † New York

Contents

Preface		**vii**
Chapter 1	Forest Certification Programs *Katie Hoover*	**1**
Chapter 2	Forest Management for Resilience and Adaptation *Ross W. Gorte*	**51**
Index		**85**

PREFACE

The national forests have been the focus of controversy for many years. Reduced timber harvests, increased wildfire risks, degraded forest health, and disagreements among users and other stakeholders have led to congressional disputes over appropriate management. Some interests have suggested third-party certification of sustainable management of the national forests as a possible solution to many of these difficulties. This book outlines the history and development of forest certification as a voluntary, market-based mechanism to promote sustainable forest practices. The two major certification programs are described, compared and contrasted based on a broad range of sustainable forest management issues pertinent to the National Forest System.

Chapter 1- The national forests have been the focus of controversy for many years. Reduced timber harvests, increased wildfire risks, degraded forest health, and disagreements among users and other stakeholders have led to congressional disputes over appropriate management. Some interests have suggested third-party certification of sustainable management of the national forests as a possible solution to many of these difficulties. There are two major certification programs in the United States: the Sustainable Forestry Initiative (SFI) and the Forest Stewardship Council (FSC) program.

The FSC and SFI programs are very similar in many regards. Both programs use a multi-stakeholder approach that balances environmental, social, and economic interests to negotiate broadly acceptable standards of sustainable forest management. Both programs use independent, accredited third parties to verify compliance with the standards. Both programs have stakeholder involvement and public transparency requirements. Within the standards, both programs have similar coverage in terms of requirements for harvest operations, wildlife and habitat management, water and soil protection, and decision-making and management planning.

Chapter 2- U.S. forests are primarily temperate forests, often with relatively few species dominating over wide areas. Such forests respond and adapt to an array of environmental factors—sunlight levels and duration, temperature, precipitation, and a multitude of disturbances (e.g., fires, pests, and storms)—and these factors could be further altered by long-term shifts in natural climate variability and climate change. Many domestic forests are already under stress from drought, severe wildfires, and insect epidemics. Changing conditions and disturbances could diminish the goods and services that forests provide—timber, clean water, scenic vistas, carbon sequestration, and much more. Forests can be subject to management techniques to improve their resilience and adaptability, to assure continued production of the economically desired ecosystem goods and services.

Congress provides management direction and funding for federal forests and financial and technical assistance for management of nonfederal forests. To date, such legislative direction and funding have at most indirectly encouraged management that promotes resilient forests. Congress could decide to broaden its role in promoting resilient and adaptable forests. However, thus far, no broad-based legislation has been introduced calling for direct federal forest management and nonfederal forestry assistance to sustain forests in the face of changing conditions.

In: Forest Certification and Sustainable ... ISBN: 978-1- 62081-206-8
Editors: J. G. Robles and C. A. Savage

Chapter 1

FOREST CERTIFICATION PROGRAMS*

Katie Hoover

SUMMARY

The national forests have been the focus of controversy for many years. Reduced timber harvests, increased wildfire risks, degraded forest health, and disagreements among users and other stakeholders have led to congressional disputes over appropriate management. Some interests have suggested third-party certification of sustainable management of the national forests as a possible solution to many of these difficulties. There are two major certification programs in the United States: the Sustainable Forestry Initiative (SFI) and the Forest Stewardship Council (FSC) program.

The FSC and SFI programs are very similar in many regards. Both programs use a multi-stakeholder approach that balances environmental, social, and economic interests to negotiate broadly acceptable standards of sustainable forest management. Both programs use independent, accredited third parties to verify compliance with the standards. Both programs have stakeholder involvement and public transparency requirements. Within the standards, both programs have similar coverage in terms of requirements for harvest operations, wildlife and habitat management, water and soil protection, and decision-making and management planning.

* This is an edited, reformatted and augmented version of Congressional Research Service, Publication No. R41992, dated September 8, 2011.

Despite these similarities, the SFI and FSC programs do have some distinct differences. The programs each emphasize different sustainability objectives: the SFI program emphasizes sustainable timber harvesting, and places forest management as a tool to achieve that objective; the FSC program emphasizes sustainable forest management, and places timber harvests as one tool to achieve that objective. The SFI standard is generally more flexible, while the FSC standard is generally more prescriptive with more on-the-ground performance requirements.

How certification would affect the management of the national forests is uncertain. However, certification could evaluate the extent to which the forest management plans align with the standards of each certification program, and then evaluate the extent to which those forest plans are being implemented. A third-party evaluation of the forest plans, and their implementation, could potentially alleviate—or escalate—stakeholder and congressional disputes over the appropriate management of the national forests.

It is unclear whether the Forest Service has the existing authority to certify the national forests. If Congress chooses to require certification of the national forests, there are other questions to consider, including which certification program(s) to require; what (if any) forest management process requirements (e.g., public involvement standards) might be relaxed; and what would be the impact on timber purchasers of processing certified sustainable wood. Congress may also consider if certification should occur across the entire National Forest System, or at the unit level, and then how many and which units should be certified.

INTRODUCTION

Nearly one-third of the land area of the United States is covered by forest land. The 750 million acres of U.S. forests are owned both privately—by individuals and corporations—and publicly— by federal, state, and local governments. At the federal level, the Forest Service—an agency within the U.S. Department of Agriculture—is the principal forest management agency, charged with managing 193 million acres designated mostly as national forests.[1] The Forest Service manages the forest resources under a multiple use-sustained yield mandate, meaning that the agency must balance uses such as providing recreation opportunities, timber supplies, livestock grazing, watershed protection, and fish and wildlife habitats, in a manner that does not impair the productivity of the land for future resource yields.

The national forests have been the focus of controversy for many years. Reduced timber harvests, increased wildfire risks, degraded forest health, and disagreements among users and other stakeholders have led to congressional disputes over appropriate management. Some interests have suggested third-party certification of sustainable management of the national forests as a possible solution to many of these difficulties.

Sustainable forest certification programs were created in the 1990s as a nongovernmental approach to promote sustainable forest management practices. Sustainable forest management generally requires that the stewardship and use of the forests balance current and future demand for the resources with maintaining ecological, economic, and social functionality. Forest landowners may voluntarily participate in these programs with the goal of gaining a market advantage through independent third-party assessments and product labels to communicate compliance with standards of sustainable management. The standards are developed by different multi-stakeholder groups with little or no government participation. Although initially directed at private forest owners, certification programs have evolved and developed mechanisms to certify public forests.

In 1997, the USDA Forest Service declared it would not pursue certification of any of the national forests. However, the certification of several state, Department of Energy (DOE), and Department of Defense (DOD) forests has raised new questions about this policy. In 2005, the Forest Service sponsored a study evaluating the potential certification of the national forests, but the agency has not issued any changes to its certification policy since the study was released in 2007.

If Congress determines that the national forests should be subject to certification, Congress could direct the Forest Service to pursue certification for some or all of the 105 national forest administrative units. Certifying the national forests as sustainably managed may have many potential benefits. Certification might demonstrate and emphasize that the management of the national forests is held to high ecological, economic, and social standards. Certification might reduce existing conflicts over management decisions for the national forests, and has the potential to streamline the management process. For example, certification programs have requirements for planning, management, and public involvement programs as well as for measurable performances on the ground. While these requirements currently exist in policy, regulations, or statutes, Congress may consider that certification would be sufficient and could replace the existing structure. In addition, certification of the national forests may provide benefits to the private sector. The timber and wood products industries as well as the "green" products industries may benefit from having access to a greater supply of certified materials. Finally, certifying the sustainability of the national forests may complement other sustainability policy directives, such as federal green purchasing and green building policies.

However, certifying the national forests may present some costs, risks, and uncertainties. The certification process would require direct and indirect investments. Direct costs include the initial fee to pay for the audit process, and then whatever annual or semi-annual costs are required to maintain certification. Indirect costs include changes that may be required to earn or maintain certification, such as implementing new monitoring and reporting systems or adjusting harvest levels. These costs may or may not be recovered in improved management, or even in higher stumpage prices for the certified products leaving the national forests. While certification was initially seen as a means to gain market advantages, those expectations have rarely been realized in the domestic U.S. market. In addition, the certification process and maintenance requirements could have the unintended, opposite effect of becoming another layer of administrative complexity for the Forest Service to navigate.

This report outlines the history and development of forest certification as a voluntary, market-based mechanism to promote sustainable forest practices. It then describes, compares, and contrasts the two major certification programs used in the United States based on a broad range of sustainable forest management issues pertinent to the National Forest System.

BACKGROUND

History and Development of Forest Certification Programs

Sustainable forest certification programs began essentially as outgrowths from failed international negotiations to address global forest degradation issues at the 1992 United Nations Conference on Environment and Development (UNCED), also known as the Earth Summit, in Rio De Janeiro.[2] Amid increasing concern among environmental nongovernmental organizations (NGOs) and other stakeholders about the deforestation occurring in developing tropical countries, the international community met to discuss a legally binding global forest policy. However, developing countries were opposed to any legally binding agreement, based mostly on trade concerns. Instead, the process led to the Statement of Forest Principles,[3] a nonbinding set of general economic, environmental, and development guidelines that was largely considered a disappointment in terms of setting a global forest policy.[4]

In 1995 the Montreal Process[5] convened as an international effort to define measures of sustainable forest management. Participating nations—including the United States—developed national policy-level criteria and indicators as measures of the principles of sustainable forest management. The criteria and indicators included the conservation of biological diversity; maintaining forest and ecosystem productivity; conserving soil and water resources; maintaining global carbon cycles; maintaining long-term socioeconomic benefits; and a legal and institutional framework for sustainable forest management. The Montreal Process, along with the Forest Principles, formed a framework of economic, environmental, and social definition and measure of sustainable forest management that was carried forward by the forest certification programs.

The Forest Stewardship Council (FSC) Standard

In the absence of governmental mechanisms, the NGO community turned to private-sector tools to promote sustainable forest management. In 1993, the Forest Stewardship Council (FSC) certification standard was developed by the World Wildlife Fund and other NGOs. As a voluntary certification and labeling system, the FSC standard was intended to leverage the growing consumer demand for sustainable wood products and use market incentives to promote sustainable forest operations. The FSC is registered as a nonprofit organization headquartered in Oaxaca, Mexico. Although the initial impetus for the development of the program was to protect tropical forests, FSC is a global standard that also applies to temperate and boreal forests. As such, the FSC principles of sustainable forest management apply globally, but regional, national, and even subnational standards may be created to adapt to local conditions. In 1995, the U.S. chapter of the FSC was formed in Minneapolis and is responsible for developing and publishing the U.S. standard applicable to the lower 48 states. The FSC-US is further divided into nine regions, with regionally specific guidance built into the standard.

As of July 2011, the FSC had certified 353 million acres of private and public forest land in 79 countries (Figure 1), including 33 million acres in the United States (Figure 2), as sustainably managed.[6]

The Sustainable Forestry Initiative (SFI) Standard

After 1993, other certification programs soon emerged to compete with the FSC program. In particular, programs emerged from various industry groups. In North America in 1994, the American Forests & Paper Association (AF&PA) established the Sustainable Forestry Initiative (SFI) to focus on sustainable forest production in the United States, and later expanded it to include Canada. The initial SFI program was developed entirely by wood products industry representatives, with no external stakeholder participation.[7] Participation became mandatory for AF&PA members, but firms could self-select which criteria to measure and could self-report compliance. Market pressures have forced the SFI standard to undergo significant substantive and structural changes to compete with the FSC standard.[8] The SFI standard now requires third-party audits, is independent from the AF&PA, and is a registered nonprofit organization headquartered in Washington, DC.

As of April 2011, the SFI had certified 58 million acres private and public forest land in the United States as sustainably managed (Figure 2).[9]

Other Certification Programs

Several other certification programs emerged in the 1990s, but competition and consolidation left two global programs—the FSC and its primary global competitor, the Programme for the Endorsement of Forest Certification (PEFC). Numerous other programs operate at national or regional levels. In North America, there are two national programs—one for Canada and the SFI for North America. Smaller programs, such as the American Tree Farm System, also operate in the United States. However, the SFI and FSC programs are the two major certification programs that would apply to the national forests.

The PEFC program is a global umbrella organization for different national certification programs, providing a common label to facilitate trade in a global market. The PEFC is a nonprofit organization headquartered in Geneva, Switzerland. Originating in Europe, the PEFC was first initiated by landowner groups, and now also includes industry interest groups. The PEFC endorses national programs using a third-party verification process based on internationally developed sustainability benchmarks. In 2005, the PEFC endorsed the SFI standard, and the SFI Board of Directors serves as the governing body for the U.S. chapter of the PEFC. As of July 2011, the PEFC had 35 national members, and had certified the sustainable management of 576 million acres (Figure 1) through the endorsement of 30 national certification programs.[10] Since the PEFC program works by endorsing other national certification programs, the PEFC program would not apply directly to the national forests.

The other national system operating in North America is the Canadian Standards Association Sustainable Forest Management System (CSA-SFM). Developed in 1996 at the request of the Canadian forest products industry association, the CSA-SFM is Canada's official national standard for sustainable forest management and is endorsed by the PEFC. As of September 2010, the CSA-SFM had certified over 160 million acres in Canada.[11] The CSA-SFM does not apply to forests in the United States.

The American Tree Farm System (ATFS) promotes responsible forest stewardship on smaller, privately owned, nonindustrial forests in the United States. Founded in 1941, the ATFS is the oldest certification system and is a program of the American Forest Foundation, a nonprofit organization headquartered in Washington, D.C. Landowners with at least 10 acres of managed forest land are eligible to participate in the program. Requiring third-party verification, the ATFS is endorsed by the PEFC and accepted by the SFI chain-of-custody requirements. The ATFS has certified over 26 million acres in the United States (Figure 2).[12] As a program that targets mostly small landowners, the ATFS would not apply to the national forests.

Elements Common in Certification Programs

Most certification programs have common elements. Participation is voluntary, but compliance with the standards is mandatory to receive and maintain certification. Most programs are private and nongovernmental, and the governance structure is typically balanced between competing environmental, economic, and social interests. The process of negotiating the standards of sustainable management is also balanced between these competing stakeholder groups to prevent one interest from dominating the process. Most programs use independent, accredited third parties to verify compliance with the standards and have requirements that extend throughout the supply chain to maintain the certified label. In addition, there are monitoring and reporting requirements, as well as re-certification requirements after a specified time frame.

Market Forces

Sustainable forest certification is a form of private governance that uses the market to induce and monitor sustainable behavior. Forest operations are evaluated by independent auditors for conformance against negotiated standards of sustainability. The result is an eco-label on forest products leaving the certified forest, which communicates to consumers a responsibly and sustainably produced product. Firms along the entire supply chain must certify their operations to continue to sell the products with the eco-label (this is referred to as chain-of-custody certification).

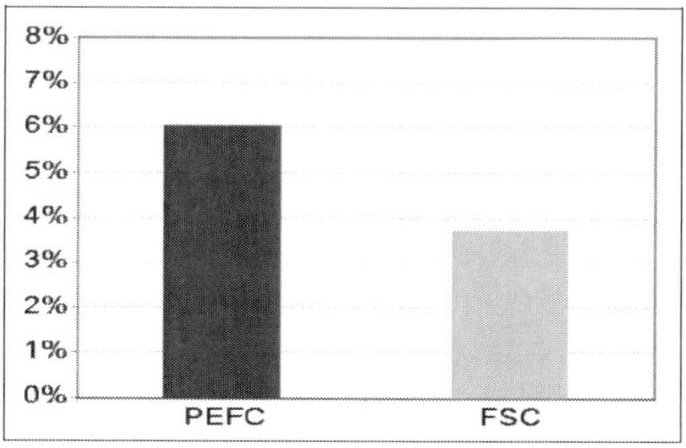

Sources: CRS, http://www.pefc.org, and http://www.fsc.org.

Notes: PEFC = Programme for the Endorsement of Forest Certification. FSC = Forest Stewardship Council.

Figure 1. Percentage of Certified Global Forested Land by Global Certification Program.

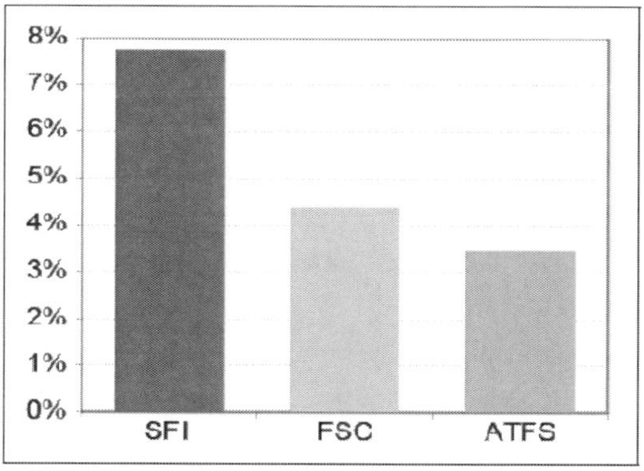

Sources: W. B. Smith et al., *Forest Resources of the United States-2007*; USDA Forest Service; http://www.sfiprogram.org,http://www.fscus.org;and http://www.tree farmsystem.org.

Notes: Includes both public and private forestland. SFI = Sustainable Forests Initiative. FSC = Forest Stewardship Council. ATFS = American Tree Farm System.

Figure 2. Percentage of Certified U.S. Forested Land by National Certification Program.

Demand by wood suppliers has driven the market for certified products more than demand by consumers.[13] For example, in the late 1990s, environmental groups concerned with rapid tropical deforestation began boycotting the Home Depot, one of the largest suppliers of wood products. In response, the Home Depot began carrying FSC-certified wood. Other retailers followed suit, creating a demand for certified wood products. Landowners began certifying their forests and providing certified products to respond to the growing demand. A similar situation occurred for pulp and paper products when Time, Inc., committed to using certified paper sources.[14] The result is that the market for certified products has grown exponentially in under two decades of existence, with nearly a quarter of the world's industrial wood production coming from certified sources.[15]

Supplier demand in the market also drives competition between the certification programs. After the boycotts, the Home Depot initially accepted only FSC-certified wood products, despite pressure from the SFI program and the AF&PA. Environmental groups—who had very successfully boycotted the Home Depot—preferred the FSC program and questioned the validity and rigor of the self-reporting and voluntary SFI program. The Home Depot commissioned a consultant to conduct an independent comparison with participation from both the SFI and FSC programs.[16] The report highlighted a stronger emphasis on ecological protections in the FSC program, and a stronger emphasis on industry concerns and a more lenient verification procedure in the SFI program. Largely in response, the SFI program went through several revisions to become more competitive and comparable to the FSC program, including requiring third-party audits and separating from the AF&PA organization. Competition has also affected the FSC program: it has revised its standard, strengthened its accreditation processes, and created programs for smaller landowners in response to the SFI and PEFC programs.

Market forces continue to influence the development and evolution of the certification programs. Since home building is the largest use of wood products, building codes are critical in determining wood use. This is especially true in the specialized green building market. The U.S. Green Building Council (USGBC) implements the Leadership in Energy and Environmental Design (LEED) program and only awards credit for FSC-certified wood products. The SFI has continued to revise its standards to become more comparable to the FSC standard, gain LEED credit, and subsequently gain greater access to the green construction market. In 2010 there was an unsuccessful movement to add other certification programs—including the SFI standard—to the LEED program. Although the initial reasoning for only allowing FSC certification is unclear,[17] the USGBC membership voted against changing the requirement.

Benefits of Certification

There are generally two reasons why landowners and firms choose to invest in forest certification programs: the potential for a market advantage, and to communicate and be accountable for responsible and sustainable forest operations.

Potential market advantages from certification include price premiums, greater access to niche markets, product differentiation, and customer loyalty. For U.S. forests, the expectations for market benefits appear to be mostly unrealized.[18] The consumer demand for certified products is limited and price premiums have been minor, and have not always offset the costs of certification.[19] Highly specialized markets, however, have realized some market advantages. Certified wood from tropical countries typically demands higher price points, as do luxury home building products (at least prior to the 2008 real estate crash).[20] Instead of just a market advantage, certification actually may be becoming a market access requirement.[21]

Despite disappointing market performances, forest landowners certified in the United States appear generally satisfied with their decision to certify. This may be because of the accountability mechanisms inherent in certification programs. Certification signals responsible management and behavior. In that sense, certification can be a defensive tool used to avoid any negative publicity similar to the Home Depot boycotts in the 1990s.

How Certification Works

Both the FSC and the SFI programs have two parts: landowner certification to ensure the sustainable management of the land, and chain-of-custody certification to trace the life cycle of wood products originating in a certified forest.

Although certification programs were established to deal with logging and timber products in the marketplace, timber harvesting is not required. Certification evaluates the extent to which forest management is meeting the objectives laid forth in a forest management plan, and then evaluates the extent to which the plan is meeting the standards of sustainable forest management.

On a broad scale, both the FSC and the SFI operate within similar frameworks with similar governance structures, procedures to develop the standards, and accreditation and auditing requirements. System governance and the development of the sustainability standards are balanced among the competing groups interested in sustainable forest management. This forces the environmental, social, and economic/industry groups to work together and negotiate standards that are broadly acceptable measures of sustainability.[22]

The certification audit is done by an accredited, independent third party that evaluates the management plan and the on-the-ground activities, and includes conversations with staff as well as with stakeholders. The audit process may find minor or major non-conformances with the standard, which must be remedied prior to or immediately following certification. Upon certification, there are maintenance and recertification requirements. Each system includes monitoring, consistency, and enforcement provisions.

Once a forest is certified, products leaving the forest may use the system's label. However, a separate, chain-of-custody certification is required in order to use the certification label along the supply chain. The chain-of-custody standard is designed to ensure that certified wood is accounted for as it leaves the forest, and typically includes sourcing, production controls, record-keeping, and documentation requirements. Each producer or vendor along the supply chain must be certified in order to use the label. Both the SFI and FSC programs use a chain-of-custody certification process that is adapted from the landowner certification process described above.

Despite their broad similarities within this general framework, the FSC and the SFI each have their own unique attributes at a detailed, specific level. Table 1 provides a side-by-side comparison.

Governance and Membership

The two certification programs operate at different levels: the FSC is an international organization with a U.S. chapter and regional distinctions; the SFI is a North American organization that is endorsed by an international certification system.

Although they both function at the national, U.S. level, the different scopes do have implications for the governance and membership structures of the programs. Both programs emphasize a balanced governance structure, but take different approaches to specific governance operations, including membership.

Table 1. FSC and SFI Structure and Governance Comparison

Comparison Criteria	FSC	SFI
Membership	Global FSC: General Assembly membership open to any individual or organization, except governments.	No members
	Members organized into 3 chambers (environmental, economic and social) with equal voting. Within each chamber, voting is divided equally between North and South members.	
	FSC-US: Membership consists of global FSC members and U.S. certificate holders.	
Board of Directors	Global FSC: Nine Board of Directors, three from each chamber elected by General Assembly	18 members, with equal representation from environmental, social and industry interests.
	FSC-US: Nine Board of Directors, three elected from each chamber from FSC-US members	Nominated & elected by current board
		Also has the External Review Panel – which is independent from SFI – as an oversight committee
Standards Review	FSC-US: Every five years	Every five years
Standards Revision Process	FSC-US: Conducted by FSC-US working groups	Conducted by a task group appointed by Board of Directors
Latest Standard Released	June 2010	January 2010
Third-Party Certification Required	Yes	Yes
	Third parties must be accredited by ASI, Inc	Third parties must be accredited to ANSI-ASQ National Accreditation Board
Certification Audit Requirements	Audit report must be peer-reviewed Final summary audit report publicly available	Final summary audit report publicly available
Non-conformance	Minor non-conformances must be resolved within three months	Minor non-conformances must be resolved within one year
	Major non-conformances must be resolved within one year	Major non-conformances must be resolved prior to certification
Standards compliance	Mandatory, very limited exceptions	Mandatory, exceptions available
Re-certification	Required every five years	Required every three years

Comparison Criteria	FSC	SFI
Dispute Resolution	Handled first by the certification body; then FSC-US; then global US	Handled first by implementation committees at the state level, then the External Review Panel

Source: http://www.sfiprogram.org/index.php and http://www.fscus.org/.

Forest Stewardship Council (FSC)

The FSC operates as a global organization[23] with a national chapter in the United States.[24] Globally, the FSC is a multi-stakeholder organization with membership open to individuals and organizations (the General Assembly), with the exception of governments. Governmental agencies may not become members, but may serve advisory or other roles.[25]

Members are organized into either the social, environmental, or economic chamber. Emphasizing a balanced approach, each chamber carries the same voting weight, regardless of membership size. Within each chamber, votes are equally divided between northern and southern hemisphere nations, with the goal of creating a balance between the needs of developed and developing nations. The General Assembly elects three members from each chamber to serve on the Board of Directors.

General Assembly meetings are held every three years, during which the membership may change the constitution, operation, and structure of the organization with a supermajority (67%) vote.[26] It is also important to note that certification and membership are separate processes; becoming certified does not automatically result in becoming a global FSC member, and not all members are certificate-holders.

FSC-US is the national initiative of the FSC charged with adapting and applying the global FSC principles and criteria to the specific conditions in the continental United States.[27] The U.S. chapter consists of a national board with nine members, which mirrors the structure of the global FSC with balanced social, environmental, and economic chambers. Global FSC members residing in the United States form the membership body and have the opportunity to participate in FSCUS committees and working groups, the standards review and revision process, and elect or may become members of either the U.S. or international Board of Directors.

Unlike the global FSC, certificate holders, including governmental agencies, automatically become FSC-US members with voting privileges.[28] Within the U.S. chapter, there are nine regional working groups.[29]

Sustainable Forestry Initiative (SFI)

Although endorsed at the international level by the PEFC, SFI operates entirely at a national level within the United States and extends operations into Canada. The SFI is not a membership-based organization but has a Board of Directors consisting of 18 members representing a balance of environmental, economic, and social interests. Current board members nominate and elect successive members; the only requirement is knowledge of sustainable forestry. The Board of Directors sets the policies of the SFI organization, elects the executive committee, appoints task groups, and develops the standards, certification, and accreditation procedures. To provide oversight, the External Review Panel—an independent group of experts representing environmental groups, professionals, academics, and public agencies—monitor the development and implementation of the SFI program. The SFI program includes Implementation Committees, which are state or regional-level grassroots organizations that promote the SFI program and sustainable forestry practices as well as respond to local needs. Although membership requirements may vary by committee, generally any SFI certificate holder with operations in that location may participate. As of August 2011, there were 37 Implementation Committees in operation at the state or regional level.[30]

Standards and Standards Development

Both certification programs use a hierarchical standard based on their particular principles of sustainable management at a broad, abstract scale and then progress towards more specific and measurable indicators.[31] The principles are defined and applied differently, largely due to the different geographic scopes of the programs. See Table 2 for a listing of the principles. Both programs emphasize an open and inclusive process to develop the standards, but there are functional differences that stem in part from the differing governance structures. Both programs released revised standards in 2010.

Forest Stewardship Council (FSC)

The global FSC standard is based on 10 principles of sustainable forest management which are implemented through 56 criteria. The FSC focuses on promoting forest stewardship, biological diversity, and ecosystem functionality while protecting the land tenure and use rights of indigenous peoples and the health and safety of forest workers. The principles and criteria

Table 2. The FSC and SFI Principles of Sustainable Forest Management

FSC Principles	SFI Principles
1. Compliance with Laws and FSC Principles	1. Sustainable Forestry
2. Tenure and Use Rights and Responsibilities	2. Forest Productivity and Health
3. Indigenous Peoples' Rights	3. Protection of Water Resources
4. Community Relations and Workers' Rights	4. Protection of Biological Diversity
5. Benefits from the Forest	5. Aesthetics and Recreation
6. Environmental Impact	6. Protection of Special Sites
7. Management Plan	7. Responsible Fiber Sourcing Practices in North America
8. Monitoring and Assessment	8. Avoidance of Controversial Sources including Illegal Logging in Offshore Fiber Sourcing
9. Maintenance of High Conservation Value Forests	9. Legal Compliance
10. Plantation Management	10. Research
	11. Training and Education
	12. Public Involvement
	13. Transparency
	14. Continual Improvement

Source: http://www.sfiprogram.org/index.php and http://www.fscus.org/.

are globally applicable, and are adapted to local conditions through 192 Indicators set at the national level by the FSC-US standard with specific guidance as needed for the 9 U.S. regions.

The global FSC is currently in the process of conducting the first full revision of the principles and criteria, although the standard has been amended to include plantations and small-scale ownerships. The General Assembly is expected to vote on the revision in late 2011.[32] However, the FSC-US standards, in accordance with the global FSC policy, are reviewed and revised every five years. The revision process is conducted by working groups of FSC members and technical experts, and includes a robust stakeholder consultation process. The FSC-US standards require global FSC approval prior to becoming effective. The latest revision was approved in July 2010. Certificate holders typically have one year to comply with a revised standard.

Sustainable Forestry Initiative (SFI)

The SFI standard is based on 14 principles of sustainable forest management which consist of 20 Objectives implemented through 38 performance measures and 115 indicators. Because the United States has a mature system of laws and regulations that have mostly settled land tenure (ownership), use rights, and rights of indigenous peoples, the SFI principles of sustainable forestry focus more on forest stewardship, biological diversity, and forest operations, although worker safety, training and improvement are also covered. Different objectives apply depending on the type of forest operation (land management or fiber sourcing) and the certification type.[33]

The SFI standard is reviewed and revised every five years. The process is conducted by a task group—with a balanced membership—appointed by the Board of Directors and includes public notice, review, and comment provisions as well as regional stakeholder workshops. In addition, the External Review Panel ensures that public input has been adequately addressed during the revision process. The latest revision became effective with Board of Directors approval in January 2010. Certificate holders typically have one year to comply with a revised standard.

Accreditation

The certification audits are not conducted directly by FSC or SFI, but are conducted by independent third-party organizations ("certification bodies") that must be internationally accredited. The accreditation process ensures these certification bodies are knowledgeable about sustainable forest management and can objectively and credibly evaluate conformance with the standards. The certification bodies may be for-profit and not-for-profit organizations. Both programs have conflict-of-interest policies to maintain the credibility and impartiality of the certification bodies. Both programs require annual audits of the certification bodies to maintain their accreditation. Many certification bodies are accredited to certify both SFI and FSC programs.

Both programs draw from the International Standards Organization[34] (ISO) 9000 and 14000 standards series. The ISO is an international standards setting organization of 146 countries and publishes standards for a variety of products and services. The ISO 9000 standard focuses on quality management process to achieve customer and regulatory requirements and includes requirements for auditors. The ISO 14000 standard focuses on the development of an environmental management process.

Forest Stewardship Council (FSC)

The FSC developed its own accreditation process and organization—Accreditation Services International (ASI)—instead of relying solely on ISO standards. Initially part of FSC, in 2006 ASI became an independent organization and currently certifies other sustainable resource management programs. In addition to ISO requirements, the FSC has specific training and education requirements for lead auditors and requirements for the composition of multidisciplinary audit teams.

Sustainable Forestry Initiative (SFI)

The SFI program requires accreditation by the American National Standards Institute (ANSI) National Accreditation Board,[35] the U.S. accreditation body for the ISO standards. The SFI program also has specific qualification requirements for the lead auditor, as well as for all members of the audit team.

Certification Audit, Monitoring and Complaints

The certification audit typically includes a review of the operating plans and procedures and other supporting documents, staff consultations, on-the-ground field inspections, and stakeholder consultations conducted by the certification body.[36] The certification body then issues a report identifying any potential minor or major non-conformances and corrective actions required to achieve certification. The final certification assessment—or a summary—is made publicly available on the system's website. Once certified, there are monitoring and recertification requirements that vary by system, as well as procedures for handling complaints.

Forest Stewardship Council (FSC)

The audit requires specific stakeholder consultations. The audit report must be peer-reviewed by at least one independent and qualified expert. Minor non-conformances must be resolved within three months of certification, major non-conformances within one year, or certification may be revoked. Annual monitoring and maintenance requirements include a surveillance audit that may be conducted on-site. Some documents must be made publicly available. Re-certification is required every five years. Compliance with all indicators is mandatory, with very limited exceptions under very specific circumstances. The FSC has a dispute resolution process (FSCSTD-01-005) to handle complaints and appeals brought against the FSC, an accredited certification body, or a certificate holder.[37] The process generally proceeds at the lowest level possible, with the certification body being the first level of response. If the dispute is not resolved, it then proceeds to the FSC-US level, and then the global FSC level.

Sustainable Forestry Initiative (SFI)

The audit requires stakeholder consultations as appropriate. Minor non-conformances must be resolved within one year of certification; major non-conformances must be resolved prior to certification. Annual monitoring and maintenance requirements include publicly available annual summary audit and progress reports. Re-certification is required every three years. Compliance with all indicators is mandatory, but the indicators may be substituted or modified based on local conditions, supported with a thorough analysis and adequate justification, and with the certification body ensuring consistency with the spirit of sustainable forestry. The SFI has a dispute resolution process to handle inquiries, complaints or challenges about inconsistent practices, the validity of a certification, or SFI label use. The process operates mostly through the state or regional Implementation Committees but may also proceed through the External Review Panel.

Policy on Public Federal Lands

The SFI does not have a policy preventing the certification of federal lands, and there are specific requirements for public lands incorporated into the standard.

The FSC standard also has some specific requirements for public lands incorporated into the standard. However, the FSC-US has a Federal Lands Policy that would apply to any federal government agency wishing to pursue certification.[38] The policy identifies three threshold standards that must be met prior to certification. The first threshold is a willing landowner (e.g. the Forest Service), the second threshold is public consensus regarding the management of the National Forest System, and the third threshold is the development of specific FSC standards applicable to the federal agency. The FSC applied the Federal Lands Policy when certifying DOD and DOE lands in 2004.[39] However, prior to issuing their certification, the FSC-US expressed concern about the ability of other federal forests to meet the threshold requirements.

Costs of Certification

There are direct and indirect costs associated with certification. Direct costs include the initial certification fee, and then may include annual maintenance fees, and after a specified time period, a re-certification fee. The initial certification fee covers the cost for the third-party audit, which includes the time and travel cost for a team site visit that may last up to a week, and time for reviewing plans, writing reports and issuing the certification. The certification cost may vary widely depending in part on the size of the land being certified; estimates range from less than ten cents per acre to hundreds of dollars per acre.[40] This initial investment tends to be easier for larger landowners to absorb. To encourage broader acceptance of certification across the landscape, both the SFI and FSC programs have created programs to help offset the costs of certification for smaller landowners. In terms of direct fees, the SFI and FSC programs appear to have comparable costs.[41]

The indirect costs of certification are the management and operation costs required to achieve and maintain certification. The certification audit may reveal operational practices that need to be changed, such as investment in different equipment, technologies, training, or management practices. Certification may require management investments, such as developing and implementing a monitoring and record-keeping program, or a product-tracking program for the chain-of-custody requirements, or non-commercial thinning to improve forest health in overstocked stands. In addition, certification may require planned harvest levels to be adjusted in both the short and long term, potentially affecting income generation.

SFI and FSC Program Comparisons

There are relatively few reports that directly and objectively compare the SFI and FSC programs, the two primary certification programs for large landowners in the United States.[42] Several reports claim objectivity, but actually were published by FSC, SFI, or one of their partners. In addition, many of the comparisons are based on previous versions of both the SFI and FSC standards and are no longer relevant.

One of the first objective comparisons was the report commissioned by the Home Depot in 1999, which included participation from representatives of each program. The comparison process and publication of that report in 2001 were a catalyst for significant changes in the SFI program.[43] In 2008, the Yale Program on Forest Policy and Governance compared the programs for the USGBC, but that comparison is based on outdated versions of the standards.[44] Most recently, the Dovetail Partnership, Inc., an environmental NGO that appears to have no ties to either program, released the only comparison based on the current standards.[45]

Among the few objective comparisons, the general consensus is that structurally the programs are now very similar. The FSC standard is more prescriptive while the SFI standard allows more flexibility. Each program has different strengths and weaknesses that relate to their different ideas about sustainability. The SFI program emphasizes sustainable timber harvesting, and places forest management as a tool to achieve that objective. The FSC program emphasizes sustainable forest management, and places timber harvests as one tool to achieve that objective. Overall, the strengths and weaknesses in each program are complemented by the strengths and weaknesses in the other program, and dual certification from both programs is relatively common, especially for public forests.

National Forests Pilot Certification Study

In 2005, the Forest Service sponsored a pilot certification study on the national forests; the Pinchot Institute for Conservation evaluated the certification process for both the FSC and SFI programs on five different national forests (see http://www.pinchot.org/gp /National_Forest_ Certification). The study included hiring accredited third parties to conduct certification audits and issue reports, and interviewing Forest Service staff about the process. The study was completed using the then-current, 2005 versions of each standard; both standards have since been updated. The results indicate that the planning and management requirements of the national forests lend themselves easily to certification. However, there may be some substantive requirements in both the SFI and FSC standards that may be of concern. This includes contractor safety training issues, monitoring of non-timber forest products (NTFPs), and conflicting old-growth provisions. For example, while the FSC standard prohibited old-growth harvests on any public lands, old-growth harvests were not prohibited in the relevant forest plans. In addition, the audits found areas—especially vegetation management and road maintenance or decommissioning—where lack of project implementation was preventing the forests from achieving their stated goals, which may affect certification.

Comparison Challenges

Directly comparing the SFI and FSC programs is challenging, mostly because the programs have different targets. The FSC program mostly uses performance-based targets, whereas the SFI program uses both performance- and systems-based targets. Performance targets are focused on the goals of the standard, or achieving some desired objective or condition, such as minimizing ecological damage. System targets are focused on developing a process to achieve the desired outcome, such as developing a program to minimize ecological damage. One useful way to understand the difference is that performance-based targets are generally evaluated in the field, while systems-based targets are generally evaluated in an office.[46]

Performance-based standards are more prescriptive, but attempt to ensure that the desired objectives are realized. Systems-based standards are often more flexible, which allows participants to adapt to their own particular situation, but may lack the mandatory mechanisms that ensure the desired objectives are actually realized.

Comparing the SFI and FSC Standards for the National Forests

Using the 2010 versions of both standards, this section evaluates both the FSC and SFI standards across a broad range of issues pertinent to the certification of the U.S. national forests. A more detailed, side-by-side comparison is provided in Appendix. The comparison uses the standards and criteria from the Montreal Process and issues relevant to public lands management.

The comparison is organized into six broad categories:

- Forest management
- Wildlife and habitat management
- Water and soil resource management
- Other forest uses and values
- Decision-making and management planning
- Miscellaneous

Overall, the SFI and FSC certification programs have similar coverage in these categories. The programs each emphasize different sustainability objectives: the SFI program emphasizes sustainable timber harvesting, and places forest management as a tool to achieve that objective; the FSC program emphasizes sustainable forest management, and places timber harvests as one tool to achieve that objective.

The SFI standard is generally more flexible, and includes more stringent worker safety requirements and provisions for continual improvement. The FSC standard is generally more prescriptive, and includes more stringent ecological protections. In addition, the FSC program is more detailed and complex, as shown by the comparative difference between the size of standard documents. The FSC standard is described in 75 pages, with references to several other policy documents. In comparison, the SFI standard is described in 14 pages. However, this level of detail may be a result of the global focus of the FSC program, as it also applies to countries that may not have comparable environmental and social regulations as the United States. Given the relative similarities between the FSC and SFI programs, what may actually drive any differences between the programs is the rigor and quality of the audit process.

Forest Management

The forest management category covers mostly silvicultural activities specifically related to the growth and cultivation of trees, such as harvest operations, roads management, fire and fuels management, insect and pest management, and economic considerations. Harvest operations within sustainable forestry balance forest productivity and economic benefits while limiting ecosystem impacts and resource damage. Sustainable harvest operations also have implications for biodiversity, depending on how the forest is regenerated. Regeneration may be done naturally by allowing the remaining trees to serve as seed sources, or regeneration may be done manually by planting saplings of the desired species.

Both require carefully planned and executed pre- and post-harvest activities such as site preparation and rehabilitation. Road-building activities should generally minimize habitat fragmentation and erosion and sedimentation concerns. Active forest management should emphasize restoring natural disturbance regimes, including wind, fire, and pest occurrences. This category also includes the definitions and harvest prescriptions for old-growth trees.

The SFI and FSC programs have many similar forest management provisions. Both programs emphasize healthy, functioning ecosystems and allow for site-specific fuel management strategies to restore natural disturbance regimes. In addition, both programs discourage the use of chemical pest control methods. Both programs have provisions to minimize the damage from road building activities. Both programs encourage stand biodiversity and the utilization of woody biomass, and discourage excessive harvest waste.

Despite these similarities, there are some differences between the forest management provisions in the FSC and SFI programs. While they both call for the calculation of a sustainable harvest level, the FSC requires a specific calculation method and requires harvest levels to remain below the projected growth level. The SFI does not define sustainable harvest levels or specify a calculation method. The FSC program has regionally specific guidance regarding the size of clearcut harvests while the SFI program has one size limit that applies nationally. The SFI program calls for specific reforestation time frames depending on the site regeneration method; the FSC does not have any specific reforestation time frames. While the FSC program prohibits harvesting old growth on public lands, the SFI program does not have any similar harvest prohibitions.

Wildlife and Habitat Management

The wildlife and habitat management category covers activities and impacts related to the entire ecosystem, such as endangered species and habitat protections, biodiversity requirements, exotic and invasive species controls, and wilderness or special places designations. As central tenets for sustainable forestry, biodiversity and forest ecosystem health and function are inter-related. Biodiversity can be evaluated across landscapes, specific sites, or within the genetic structure of the same species.[47] Increasing biodiversity helps the forest ecosystem withstand and adapt to various disturbances and maintain functionality.

Protecting endangered species and their habitats is in part related to biodiversity conservation. Preserving landscapes through wilderness or other special designations is also related to conserving biodiversity, but also reflects other cultural and spiritual values and uses for the forest ecosystem.

The SFI and FSC programs have a few, broadly similar wildlife and habitat management protections such as promoting stand and landscape biodiversity, controlling exotic and invasive species, and allowing the use of hybrids and clones.

Although both programs have species and habitat protections and special places designations, there are differences within the particular provisions. The FSC program has more stringent protections that extend to a broader classification of rare, threatened, and endangered species and habitats and encourages directed protection to ensure activities like habitat connectivity. The FSC also calls for a precautionary approach when evaluating actions that can affect those species, while the SFI does not specify a similar precautionary approach.

The FSC program has a broader definition of special places that include more stringent protection requirements. In addition, the FSC has more stringent protections for old growth by prohibiting any harvest on public lands. The SFI program encourages conserving old growth, but does not prohibit old-growth harvests. Finally, the FSC program explicitly prohibits the use of genetically modified organisms (GMOs), while the SFI program allows for their use as long as all applicable laws and regulations are followed.

Water and Soil Resource Management

The water and soil resource management category covers activities that may impair those resources, and the health and productivity of the forest ecosystem. Forest ecosystems often contain the headwaters for water bodies that serve multiple interests, such as recreation or municipal water supply sources. Within the forest ecosystem, erosion and sedimentation are the primary impairments to water quality, and are often associated with harvest operations.

Harvest operations may also impair the soil resource. Soils that have been eroded, compacted, or depleted of nutrients may impair the ability of the soil to support vegetation and degrade the productivity of the ecosystem.

The water and soil resource management provisions with the SFI and FSC programs are fairly similar. Harvest operations within both programs are to meet or exceed the best management practices (BMPs) for erosion and sedimentation control and water quality specified for that region. Both programs have protections and guidelines for water bodies, including riparian zones and wetlands.

The primary difference between the programs water and soil resource management provisions is that the FSC explicitly calls for the consideration of soil productivity when planning multiple harvests on the same site, and the SFI is silent on this topic.

Other Forest Uses and Values

The other forest uses and values category covers issues mostly related to the multiple-use mandate of the national forests. The Multiple Use-Sustained Yield Act of 1960 (P.L. 86-517) directed the Forest Service to manage the national forests for the uses of range (livestock grazing), recreation, timber, wildlife, and water supply. While these apply to surface use, there is also interest in the subsurface use of the national forest land, which is typically managed and administered in conjunction with other federal agencies. Also covered in this category are aesthetic and visual impacts and climate change concerns related to the forest ecosystem's role in the carbon cycle.

The FSC and SFI programs treat other forest uses and values differently. The SFI program has more prescriptive requirements for mitigating visual and aesthetic impacts from harvest operations, but is largely silent on grazing and mineral development. The FSC program has more prescriptive requirements for grazing, mineral development, and carbon storage.

For grazing and mineral development, the FSC calls for those uses to minimize ecological damage. However, the FSC's focus on land tenure and user rights tends to discourage landowner infringement on use rights associated with mineral development and grazing leases. While the SFI program calls for increasing research and knowledge about climate change impacts, the FSC standard is more prescriptive and essentially calls for no net loss of carbon storage capacity.

Decision-Making and Management Planning

The decision-making and management planning category covers activities related to the legal and institutional framework for sustainable forestry, as well as stakeholder input and public transparency issues. Sustainable forestry, in many ways, relies on careful planning to be successful. Management planning includes assessing the current conditions of various resources, understanding the implications different management options will have on the future conditions of those resources, selecting a management option that achieves the desired future conditions, and then monitoring the implementation of the management option and its impacts.

While public involvement and transparency are important considerations for the management of the national forests as a public resource, it is also an important consideration for certification programs as market mechanisms. Planning, public involvement, and decision-making for the national forests are governed by several intersecting laws, including the National Forest Management Act (P.L. 94-588), the National Environmental Policy Act (P.L. 91-190), and the Administrative Procedures Act (P.L. 79-404).

The decision-making and management planning provisions within the SFI and FSC programs are similar in many regards. Both programs call for compliance with all laws and regulations. Both programs call for the development of management plans. The management plans include the collection of a range of data and maps, and specify target harvest areas and levels. Both programs have annual reporting requirements that include making a summary report available to the public.

Despite these similarities, there are some differences within the decision-making and management planning provisions with the two programs. The FSC has a more stringent requirement for updating the management plans. The SFI has a more stringent worker safety requirement that extends to contractors.

The remaining differences, although distinct when comparing the two programs together, lose their distinctions when applied to the national forests and the mandates of various U.S. laws and regulations.

Miscellaneous

The miscellaneous category captures areas that are not relevant to the other categories but may still be of interest to Congress. Included in this category are the requirement from both certification programs for an official commitment to the program and requirements for public federal land standard compliance. Specific chain-of-custody requirements are not covered in this report.

Observations

Although initially directed at private forest owners, forest certification programs have become increasingly used by public forest management agencies. In the United States, several state and county forests have become certified, including but not limited to the Michigan Department of Natural Resources, the Minnesota Department of Natural Resources, the New York Department of Environmental Conservation, the Pennsylvania Department of Conservation and Natural Resources, the Wisconsin County Forests and the Wisconsin Department of Natural Resources, and the Washington Department of Natural Resources.

Instead of choosing just one program, most of these public forests opted for dual certification and became certified by both the SFI and FSC programs. There are no standard metrics to evaluate the benefits and costs of the state certification programs.

While the SFI program has stated that it is ready to certify the national forests, the FSC program is likely several years away from having procedures that would address a national forest certification process. Prior to certification, the FSC program would need to determine if the national forests meet the three threshold requirements set forth in its Federal Lands Policy. Then, if the national forests meet the threshold requirements, the FSC would have to develop and approve a standard specific for the national forests.

It is not clear how certification would affect the management of the national forests. However, certification could evaluate the extent to which the forest management plans align with the standards of each certification program, and then evaluate the extent to which those forest plans are being implemented.

The results of these third-party evaluations of the forest plans and their implementation could potentially alleviate—or escalate—stakeholder and congressional disputes over the appropriate management of the national forests.

It is unclear whether the Forest Service has the existing authority to certify the national forests. If Congress chooses to require certification of the national forests, there are other questions to consider, including which certification program(s) to require; what (if any) forest management process requirements (e.g., public involvement standards) might be relaxed; and what would be the impact on timber purchasers of processing certified sustainable wood.

Congress may also consider if certification should occur across the entire National Forest System, or at the unit level, and then how many and which units should be certified.

APPENDIX. FSC AND SFI STANDARDS COMPARISON

This table presents a side-by-side comparison of the FSC and SFI standards that would be applicable to U.S. national forests. The comparison criteria categories used were developed by CRS for assessing sustainable forestry, the standards and criteria from the Montreal Process, public forest land management, and other factors. Summaries of the FSC and SFI measures are presented.

Several measures are listed more than once if they are applicable to more than one area.

However, not all of the SFI and FSC measures are included in the table, as some are not relevant for the sustainable management of federal forests.

Each entry identifies the relevant FSC and SFI measures using the numbering system of that program.

Table A-1. FSC and SFI Standards Comparison

Comparison Category	FSC	SFI	Notes
Forest Management			
Harvest Operations	Select harvest operations to minimize adverse environmental and social impacts. Implement written guidelines that meet or exceed established Best Management Practices (BMPs) for site preparation, harvest prescriptions, techniques, timing and equipment. 5.6.c, 6.5.a-c	n/a	FSC places timber harvesting within larger framework of sustainable forest management. SFI places forest management as tool to achieve sustainable timber harvests.
Sustainable Harvest Level	Harvest levels do not exceed growth over successive harvests. Calculate and document the sustained yield harvest level for each harvest area. Rolling 10-year average annual harvest levels cannot exceed the calculated sustained yield level. 5.6.a-b	Recommend sustainable harvest levels for areas available to harvest, document annual harvest trends, and periodically update forest inventory and recalculate harvest levels to account for productivity changes. 1.1.1.g, 1.1.2, 1.1.4	The SFI does not define sustainable harvest levels.
Harvest Waste	Employ management practices to minimize the loss and/or waste of harvested forest products. 5.3.a	Develop a program to minimize waste and ensure efficient use of harvested trees. May include provisions such as exploring alternative use and markets for low-grade materials. Encourages management of harvest residue in respect to economic, environmental and social factors. 7.1.1	Both programs discourage harvest waste; SFI acknowledges the economic potential of harvest residue.

Comparison Category	FSC	SFI	Notes
Roads— Transportation system for harvest operations	Design, construct, maintain and/or reconstruct roads – permanent and temporary – to minimize short- and long-term environmental impacts. Extends to skid trails, water crossings and landings. Close and rehabilitate unneeded roads. Limit access to minimize environmental impacts. Meet or exceed BMPs. Roads allowed in limited and specific circumstances in some protected areas. (Representative Sample Areas, see Habitat Protection, below, for more information.) 6.5.d, 6.5.b, 6.4.c	Design and construct roads to minimize impacts to soil productivity and water quality. (BMPs for water quality are discussed below.) Minimize skidding and rutting. 2.3.4, 2.3.7	Both programs call for road planning and construction to be done in a manner that minimizes impacts. FSC calls for the closure of unneeded roads.
Clearcut Harvesting	Clearcut requirements with regionally specific opening sizes for even and uneven-aged harvests, ranging from 2 to 80 acres to no limit. Minor deviations are permitted under limited situations and with detailed site-specific planning and analysis requirements. 6.3.g.1-2, Appendix C	Average clearcut size is not to exceed 120 acres, with documentation requirements on the process of calculating the size. Exceptions allowed for forest health concerns, other natural catastrophes, and to meet regulatory requirements. Additional restrictions have minimum tree age and size requirement (three years old or five feet high) for areas adjacent to a clearcut harvest. 5.2.1-2, 5.3.1-3	The FSC indicator is based more on ecological characteristics that account for the responses of different forest ecosystems, cover types and natural disturbance regimes. The SFI clearcut requirements are based more on visual impacts than ecosystem responses.
Old-Growth Harvests	On public lands, all old growth is protected from harvesting. 6.3.a.1, 6.3.a.3, 9.1, Definitions, Appendix F	n/a	FSC prohibits the harvesting of old growth on public lands. Although the SFI calls for the conservation of old growth at a landscape level, there is no specific harvesting guidance.

Table A-1. (Continued)

Comparison Category	FSC	SFI	Notes
			The Forest Service definition of old growth is regionally/ecologically specific, which would render the effect of this FSC indicator uncertain.
Site Preparation and Rehabilitation	Use harvesting operations and management activities, including site preparation, harvest prescriptions, techniques, timing and equipment, to protect soil and water resources, avoid erosion, landslides, and significant soil disturbance, and to protect residual trees and other forest resources. Meet or exceed BMPs. 5.3.b, 6.5.b-c	Develop criteria for harvesting operations and site preparation activities to protect soil productivity, retain vigorous trees as applicable, and minimize impacts to maintain site productivity. 2.3.4-6	Both programs have similar provisions, although the FSC is more prescriptive.
Reforestation	Promote natural regeneration methods as a practice to address species diversity. Prefer local sources and native species over non-local sources when planting is required for regeneration. 6.3.e-d	Designate harvest areas for either natural regeneration or planting. Post-harvest deadlines include two years for planting or five years for natural regeneration, with flexibility for environmental or legal reasons. Consider using different species and mixes, and minimize planting exotic species unless research documents minimal risk. 2.1.1-7	SFI specifies time frames for reforestation. FSC does not specify reforestation time frames and mostly addresses reforestation through biodiversity requirements.

Comparison Category	FSC	SFI	Notes
Stand structure	Maintain, enhance or restore under-represented successional stages, habitat components and associated structures, including large live trees, decaying trees, snags, legacy trees, vertical and horizontal complexity. 6.3.a, 6.3.f	Develop and implement criteria to retain stand-level wildlife habitat elements such as snags, stumps, mast trees, downed woody debris, den trees and nest trees. Assess and account for in planning and management activities, where feasible, the different forest cover types, age or size class and habitats. 4.1.4, 4.1.5	Very similar coverage of habitat elements and site-specific structures, but FSC is more specific at the landscape-level by requiring protection of different and/or under represented successional stages and includes protections for old growth areas.
Biomass and Non-Timber Forest Products (NTFPs)	Calculate a sustained yield harvest when NTFPs are harvested in significant commercial operations. Diversify the economic base of public forests among several uses, including (but not limited to): lesser-used species of trees, grades of logs, and lumber; NTFPs; and emerging markets in new commodities. 5.4.b, 5.6.d	Explore alternative markets for underutilized species and low-grade wood. 7.1.1.d	Both encourage utilization of woody biomass; FSC also addresses NTFPs.
Fire and Fuels Management	Site-specific fuels management practices allowed, based on natural fire regimes, risk of wildfire, potential economic losses, public safety, and applicable laws and regulations. Fuel and fire management activities permitted in limited and specific circumstances in some protected areas (Representative Sample Areas, see Habitat Protection, below, for more information). 6.3.i, 6.4.c	Manage to promote healthy and productive forests and protect from environmentally and economically damaging agents, including wildfires. Develop a program to incorporate the role of prescribed or natural fire where appropriate. Participate in fire prevention programs. 2.4.1-3, 4.1.8	Both programs appear to have sufficient decision-space to create site-specific fuel management strategies.

Table A-1. (Continued)

Comparison Category	FSC	SFI	Notes
Insect and Pest Management Pesticide/Insecticide/ Herbicide/Biological Control Agents Use	Promote integrated pest management and other strategies to minimize the use of chemicals. Analyze, develop and implement strategies to promote the use of non-chemical methods over chemical methods. Use the least harmful formulation and application methods to minimize risks to non-target species and sites, specifically related to the choice between aerial and ground application. Monitor and keep strict records. Storage, transportation, disposal and spill response requirements. Use of biological control agents is allowed when supported by scientific evidence and is strictly controlled with documentation and monitoring requirements. Products on the FSC list of Highly Hazardous Pesticides are banned. Pest management activities permitted in limited and specific circumstances in some protected areas (Representative Sample Areas, see Habitat Protection, below, for more information). 6.6.a-e, 6.7.a-c, 6.8.a-c, 6.4.c, definitions, FSC Pesticides Policy 2005a	Manage to promote healthy and productive forests and protect from environmentally and economically damaging agents, including pests. Participate in pest prevention programs. Minimize chemical use required to achieve management objectives and use integrated pest management strategies where feasible. Chemicals used should be the least harmful necessary, for the intended use and in accordance with label requirements, and applied by trained or certified applicators. 2.2.1-2.2.6, 2.4.1-3	Both programs discourage the use of chemical pest control methods. The SFI indicator does not provide specific guidance about biological control agents. FSC is probably more detailed because use of such control agents is generally covered in U.S. environmental laws. Analysis of chemical versus non-chemical use is covered in the NEPA analysis for federal lands.
Economics	Responses to short-term financial factors, such as deferring road maintenance projects, are limited to levels that are consistent with fulfillment of this standard. On public lands, provisions are made to allow small businesses to bid competitively. The economic base of public forests should be diversified among several uses. 5.1.a-b, 5.2.c, 5.4.b	The only direct reference to economics is to protect forests from damaging agents to maintain and improve long-term forest health, productivity and economic viability. Several indicators call for the consideration of economic factors. 2.4	The FSC has a more stringent standard that does not allow management decisions to be changed in response to short-term financial concerns.

Comparison Category	FSC	SFI	Notes
Wildlife and Habitat Management			
Species Protection	Assess, locate, verify, and protect rare, threatened, and endangered (RTE) species that are federally or state protected. Conduct field surveys prior to site-disturbing activities to verify presence or absence of species. Adopt a precautionary management approach that assumes presence until absence is proven. Modify management as necessary to maintain, restore or enhance the extent, quality, and viability of the species and their habitats. Establish conservation zones or protected areas, with active management permitted as appropriate, to maintain or improve species viability. On public forests, design forest management plans and operations to meet species' recovery goals, as well as landscape level biodiversity conservation goals. Hunting, fishing, trapping, collecting and other activities are controlled to avoid the risk of impacts to vulnerable species and communities. 6.1.a, 6.2.a-d	Develop a program to promote conservation of native biological diversity, including species, wildlife habitats and ecological community types. Develop a program to protect federally and state threatened or endangered listed species (G1-G2, S1-S2 species). Develop a program to locate and protect known sites of critically imperiled or imperiled (G1 and G2) species, also known as Forests with Exceptional Conservation Value (FECV). 4.1.1-3	The FSC has a more stringent standard for species protection than the SFI, first by extending protections to species that are listed as rare, then by requiring the maintenance, restoration or enhancement of the quality and viability of the species and their habitats, and by requiring a precautionary approach that directs management to assume the species is present until proven absent. The stringency may be a result of the relatively strong protection of the Endangered Species Act (ESA, P.L. 93-205) in the United States, but it appears to go beyond the ESA requirements.
Habitat Protection	Establish conservation zones or protected areas for rare ecological communities or those that support RTE species, with active management permitted as appropriate, to maintain or improve habitat conditions and species or community viability. Modify management as necessary.	Program to promote conservation of native biological diversity, including species, wildlife habitats and ecological community types.	The FSC standard is more stringent than the SFI on habitat protection, first by extending protection for rare species habitat, while the SFI protects threatened and endangered species habitat.

Table A-1. (Continued)

Comparison Category	FSC	SFI	Notes
	Designate Representative Sample Areas (RSAs) to establish or maintain ecological reference conditions, create or maintain under represented ecological conditions, or serve as a set of protected areas for species, communities, and community types not captured in other criteria. Management activities within RSAs are limited. Establish and maintain a network of protected areas sufficient in size to maintain species dependent on interior core habitats on large, contiguous public forests. Minimize impacts of transportation programs on wildlife habitat, migration corridors, and habitat fragmentation. Additional habitat protection provisions for High Conservation Value Forests (HCVFs) included in the Wilderness/Special Places section below. 6.2.b, 6.3.a.2, 6.3.b, 6.4.a-c, 6.4.e, 6.5.d	Program to locate and protect known sites of critically imperiled or imperiled (G1 and G2) species, also known as Forests with Exceptional Conservation Value (FECV). Collect information on FECV and other biodiversity data through forest inventory processes and have a program or methodology to incorporate forest inventory and research results into forest management decisions to manage wildlife habitat and contribute to the conservation of biological diversity. Identify, map, catalog and appropriately manage ecologically, geologically or culturally unique special sites. 4.1.1, 4.1.3, 4.2.1-2, 6.1.1-2	The FSC allows some management activities in protected areas. The FSC also has requirements for protecting underrepresented ecosystems and ensuring habitat connectivity.
Biological Diversity	Use an integrated approach to manage for genetic, species, and ecosystem biological diversity at the stand and landscape levels. At the stand level, promote plant species diversity, maintain a full range of tree sizes and stand-level habitat elements, use local seed sources, control invasive species, and promote natural cycles through fuels management as necessary.	Develop a program to promote conservation of native biological diversity, including species, wildlife habitats and ecological community types at the stand and landscape levels. Develop and implement criteria to retain stand-level wildlife habitat elements.	Both programs promote biological diversity at the stand and landscape levels, although the FSC program is more prescriptive.

Comparison Category	FSC	SFI	Notes
	Maintain, enhance, and/or restore underrepresented successional stages, rare ecological communities, and riparian management zones at the landscape level. 6.2.c, 6.3.a-i	Collect biodiversity data through forest inventory processes and have a program or methodology to incorporate forest inventory and research results into forest management decisions to manage wildlife habitat and contribute to the conservation of biological diversity. 4.1.1-4, 4.2.2	
Genetically Modified Organisms (GMOs)	The use of GMOs is strictly prohibited, but the use of genetically improved organisms, including hybrids and clones, is allowed. 6.8.d, FSC GMO Policy[b]	Research on genetically engineered trees must adhere to all applicable federal, state, regulations and international protocols. Program for the appropriate research, testing, evaluation and deployment of improved planting stock, including varietal seedlings (hybrids and clones). 15.1.2, 2.5.1	While the FSC prohibits the use of GMOs, the SFI states that research must follow all laws, thus potentially allowing for the use of GMOs. Both programs allow for the use of varietal seedlings.
Exotic and Invasive Species	Develop and implement a strategy to prevent or control invasive species. Use exotic species, including seed mixes for erosion control, only when scientific evidence indicates a low risk to native biodiversity. Monitor and address impacts. 6.3.h, 6.9.a-c	Minimize the use of exotics during reforestation, unless scientific evidence documents minimal risk to native biodiversity. Manage to promote healthy and productive forests and protect from environmentally and economically damaging agents such as invasive species. Participate in programs to limit the introduction, impact and spread of invasive exotic species. 2.1.4, 2.4.1-2, 4.1.7	Both programs call for the protection from and eradication of invasive exotics, and minimizing the use of exotics except when their use is documented as low-risk.

Table A-1. (Continued)

Comparison Category	FSC	SFI	Notes
Wilderness and/or Special Places	Defines High Conservation Value Forests (HCVFs) in part as containing globally, regionally, or nationally significant ecosystems, species, or other biodiversity values or providing critical services to local communities. Guidance is regionally specific, but old growth, primary forests, and roadless areas greater than 500 acres are included as HCVF areas. HCVFs are managed to protect and maintain their high conservation value attributes. Permitted management activities vary by HCVF objective, but management decisions shall always follow a precautionary approach to protect conservation values. Identify, delineate, and designate HCVFs, with stakeholder input and public review. Management activities maintain or enhance the values and extent of the HCVF, with the necessary strategies described in the management and operation plans. Active management may or may not be appropriate, depending on the values of the HCVF. Requires the consideration of HCVFs adjacent to and beyond the boundaries of ownership. Monitoring requirements with an adaptive management strategy. 9.1.a, 9.2a-b, 9.3.a-c, 9.4.a-b, Definitions, Appendix F	Defines Forests with Exceptional Conservation Value (FECV) as having critically imperiled and imperiled species and ecological communities. Develop a program to locate and protect FECV sites with viable occurrences of critically imperiled and imperiled species and communities. Collection of information on FECV and other biodiversity related data through forest inventory process, mapping or participation in external programs. Such participation may include providing non-proprietary scientific information, time and assistance by staff, or in-kind or direct financial support. Identify, map, catalog, and appropriately manage ecologically, geologically or culturally unique special sites. 4.1.3, 4.2.1, 6.1.1-2	Areas designated wilderness within the national forests would be considered HCVF under the FSC standard, and may be considered FECV under the SFI standard. These extend protections beyond designated wilderness areas, if they fit the HCVF or FECV criteria and process. However, note that HCVF or FECV designation does not mean the same as a wilderness designation. FSC has a much broader and more inclusive definition. The SFI special sites section is sufficiently open to include many of the same areas, but their inclusion is not mandatory. It's not clear how the wilderness designation process fits with the FSC HCVF requirements. In the United States, wilderness can only be designated by an act of Congress.

Comparison Category	FSC	SFI	Notes
			This may be an area addressed by the FSC if/when they choose to develop a standard specific for the national forests.
Old Growth	Specifies two types and levels of protection for old growth. However, on public lands, all old growth is protected from harvesting and other timber management activities with some exceptions for restoration activities. Old-growth areas are normally designated as HCVFs. When old-growth communities are underrepresented in the landscape, a portion of the forest is managed to enhance and/or restore old-growth characteristics. Type 1: $\geq$3 acres that have never been logged and display old growth characteristics. Protected from harvesting and road construction. Type 2: $\geq$20 acres that have been logged, but which retain significant old growth structure and functions. Protected from harvesting. 6.3.a.1, 6.3.a.3, 9.1, Definitions, Appendix F	Support of and participation in plans or programs for the conservation of old-growth forests in the region of ownership. 4.1.6	The FSC has more stringent protections for old growth. The SFI encourages the conservation of old growth on a landscape level, on neighboring or nearby forests, without specifying protections for old growth on the certified forest. FSC prohibits the harvesting of old growth except in limited situations. Old-growth areas are protected, either as HCVFs or RSAs, with some management activities permitted as necessary. The Forest Service definition of old growth is regionally/ecologically specific, rendering the effect of this FSC indicator uncertain.

Table A-1. (Continued)

Comparison Category	FSC	SFI	Notes
Water and Soil Resource Management			
General	Develop and implement management approaches and field prescriptions that avoid or minimize negative short- and long-term impacts, and maintain or enhance the long-term ecological viability of the forest. Develop and implement appropriate measures for maintaining and/or enhancing forest services and resources that serve public values. Management operations should not have a significant, long-term damage to these services and resources. Measures should be taken to restore any adverse impacts resulting from past management. 6.1.c, 5.5.a-b	n/a	FSC has a policy to avoid or minimize impacts to ecological functions in general and those that provide public services. SFI does not have the same stated policy but has indicators to the same effect.
Water Quality	Prepare and implement written guidelines – including to meet or exceed all BMPs – for water quality, erosion control, road construction and all other mechanical disturbances. Assess water resources and hydrologic functions. Management activities are selected and used to protect water resources. 6.5.a-c, 6.1.a	Meet or exceed all water quality laws and U.S. EPA approved water quality BMPs. Program to implement and monitor BMPs, with contractor provisions. Plans address wet-weather events and operating conditions. Develop appropriate protection measures where BMPs do not exist. 3.1.1-4, 3.2.5	Both programs require that water quality BMPs are met.
Riparian Zones	Implement regionally specific Streamside Management Zone (SMZ) buffer guidelines that meet or exceed BMPs, including protected vegetative buffer areas and guidance for harvest and road construction. Minor variations permitted.	Develop a program addressing management and implementing protection of rivers, streams, lakes and other water bodies and riparian zones.	FSC has specific guidance directed at harvest operation activities.

Comparison Category	FSC	SFI	Notes
	Crossings are avoided when possible and, when necessary, do not impede the movement of aquatic species and temporary crossings are restored. 6.5.e.1-2, 6.5.f	Map and identify water bodies as specified in BMPs and where appropriate. Develop appropriate protection measures where BMPs do not exist. 3.2.1-5	
Wetlands	Wetlands protections are included in the SMZ (riparian zone) guidelines. 6.5.e.1-2, 6.5.f	Identify and protect non-forested wetlands of ecological significance. 3.2.4	Both programs have protections for wetlands.
Erosion and Sedimentation Control	Management activities – including the transportation system – are selected and used to protect soil and water resources and to avoid erosion, landslides, and significant soil disturbance. Regionally specific guidance for slopes and landslide prevention, including no logging activities allowed on areas at high risk for landslides. 6.5.a-d	Use erosion control measures to minimize the loss of soil and soil productivity. 2.3.3	Both programs call for erosion and sedimentation control, although the FSC program is more prescriptive and includes regionally specific guidance.
Soil Quality/Productivity	Assess soil resources. Management activities are selected and used to protect soil and water resources and to avoid erosion, landslides, and significant soil disturbance. To protect soil productivity, function and habitat, topsoil disturbance is limited, rutting and compaction is minimized and other issues such as multiple harvest rotations are addressed. 6.1.a, 6.5.c	Develop criteria that address harvesting, site preparation and post-harvest conditions to protect soil productivity. This includes a process to identify vulnerable soils, design and construct transportation system to minimize rutting and skid trails, and retain vigorous trees during partial harvesting. Use soil maps where available. 2.3.1-7	The FSC directly calls for the consideration of soil quality over multiple harvest rotations; this is captured indirectly in the SFI program through considerations of site productivity changes (1.1.1.4). The SFI does not say anything directly about landslides like the FSC does, but the language is sufficiently broad to allow for similar landslide protections.

Table A-1. (Continued)

Comparison Category	FSC	SFI	Notes
Other Forest Uses and Values			
General	Identify, define, and implement appropriate measures for maintaining and/or enhancing forest services and resources that serve public values, including municipal watersheds, fisheries, carbon storage and sequestration, recreation and tourism. Management operations should not have significant, long-term impacts on these services and resources. Measures should be taken to restore any damage resulting from past management. 5.5.a-b	n/a	
Livestock Grazing	Allows for tenure and use rights, as established by law or regulation, to continue and management activities should not impact the benefits of such rights. When there is a conflict between the use of such rights and the conservation of forest resources, it is brought to the attention of the certifying body. Livestock grazing is controlled to protect in-stream habitats, water quality, riparian vegetation, and the banks of the stream channel from erosion. Location, intensity and season of use should be managed to avoid adverse impacts. 2.2.a-b, 6.5.h	n/a	In the FSC program, livestock grazing is not specified as a use right, although it should fall in this category as a long-term lease. The use rights provisions in the FSC are intended to protect the rights of indigenous people, but as written appear to apply to long-term leaseholders such as those for livestock grazing and mineral development. This may be an area addressed by the FSC if they choose to develop a standard specific for the national forests.

Comparison Category	FSC	SFI	Notes
Minerals Development & Extraction	Allows for tenure and use rights, as established by law or regulation, to continue and management activities should not impact the benefits of such rights. When there is a conflict between the use of such rights and the conservation of forest resources it is brought to the attention of the certifying body. Identify areas and facilities associated with subsurface mineral and gas rights. Control the surface disturbances to the extent allowed to minimize adverse environmental and social impacts. 2.2.a-b, 6.10.f	n/a	It is not clear how the FSC would treat the separate ownership of sub-surface mineral rights (split estates). This was identified as an issue for each of the five national forests in the pilot certification study. This may be an area addressed by the FSC if/when they choose to develop a standard specific for the national forests. On federal lands, administration of mineral exploration and development is complicated by the involvement of many federal agencies, including but not limited to the land management agency.
Recreation	Recreation use is managed to avoid damage to soils, water, plants, wildlife and wildlife habitats. Recreational trails and off-road travel are controlled to minimize impacts. 6.5.d, 6.5.g	Support and promote recreational opportunities for the public. 5.4.1	FSC has more prescriptive requirements for minimizing the damage to ecological functions from recreation.
Visual Impacts	Incorporate an understanding of the likely social impacts, including aesthetic considerations, of management activities into management planning and operations. 4.4.a	Develop a program to address visual quality management that incorporates aesthetic considerations in harvest, road, and landing design and management, and other management activities where visual impacts are a concern.	SFI has more prescriptive requirements for visual and aesthetic impacts.

Table A-1. (Continued)

Comparison Category	FSC	SFI	Notes
		Also to address visual concerns, areas adjacent to a clearcut harvest must meet a minimum tree age and size requirement (three years old or five feet high). 5.1.1-2, 5.3.1-3	
Climate Change/Carbon Sequestration	Identify, define and implement appropriate measures for maintaining and/or enhancing forest services and resources, including carbon storage and sequestration. 5.5.a-b	Monitor information from regional climate models on long-term forest health, productivity and economic viability. Gain knowledge about climate change impacts on wildlife, wildlife habitats and conservation of biological diversity. 15.3.1-2	The FSC standard is more prescriptive.
Decision-Making and Management Planning			
Regulatory Compliance	Comply with all federal, state, and local laws and regulations, including relevant provisions of all applicable binding international agreements. Ensure that employees and contractors are duly informed about applicable laws and regulations. Meet or exceed all applicable laws and regulations covering the health and safety of employees and their families. 1.1.a-b, 1.2.a, 1.3.a, 4.2.a-b	Comply with applicable federal, state, and local laws and regulations pertaining to forestry and related social issues, as well as international protocols related to research on genetically engineered tress. Meet or exceed all water quality laws. 14.1.1-3, 14.2.1-2, 15.1.2, 3.1	The FSC considers the number, severity and temporal pattern of legal/regulatory violations and complaints for the five years prior to certification assessment. If laws prevent compliance, not obtaining certification is possible.

Comparison Category	FSC	SFI	Notes
Management Plans	Develop a management plan that describes historic, current and desired future conditions and includes the collection of a range of ecological, aesthetic, cultural, and socioeconomic data and maps. The management plan also describes management objectives required to achieve desired future conditions, monitoring requirements and stakeholder involvement. The management plan is reviewed and updated as necessary, but at least every 10 years. Significant plan or land ownership changes are disclosed to the FSC within 90 days. 7.1.a-r, 7.2.a, 6.4.a-b,1.6.c	Develop a forest management plan that includes the collection of a range of ecological data, such as a forest inventory, long-term resource analyses, land classifications, soils inventories and maps, growth-and-yield models, recommended harvest levels, and a review of non-timber issues. 1.1.1, 4.1.5	Both programs have management planning requirements that fall within the mandated land and resource management plan requirements of National Forest Management Act (NFMA, P.L. 94-588). The FSC requirements include the collection and assessment of a comprehensive range of ecological and social issues; whereas the SFI management planning requirements are mostly focused around ecological issues. The FSC management plan requires updates every 10 years. NFMA mandates plan updates every 10-15 years, although some updates have taken longer. The SFI does not specify plan updates or a time frame for their completion.
Sustainable Harvest Plans	Calculate the sustained yield harvest level for each planning unit and consider the impact of repeated harvests. Rolling 10-year average annual harvest levels cannot exceed the calculated sustained yield level. 5.6.a-b	Recommend sustainable harvest levels in the management plan and document annual harvest trends in relation to plans. 1.1.1.g, 1.1.2	The FSC has a more stringent requirement.

Table A-1. (Continued)

Comparison Category	FSC	SFI	Notes
Impact Assessment	Analyze the potential short- and long-term social and environmental impacts of management actions prior to conducting any site-disturbing activities. Management actions and field prescriptions are developed and implemented that avoid or minimize possible damage. 4.4.a, 6.1.a-d	n/a	This is mostly covered by the impact assessment requirements of NEPA. One area of concern is that the FSC standard requires avoiding or minimizing damage, which is not required by NEPA.
Implementation, Monitoring & Evaluation	Develop and implement a regular and comprehensive monitoring protocol and record-keeping system to evaluate various forest statistics, the environmental and social impacts of harvest and management operations, the impacts of the transportation system, stakeholder responses, and the costs, productivity and efficiency of forest management. Evaluate the extent site-specific plans achieve management objectives and minimize environmental impacts. Evaluate the extent the management plan is achieving management objectives and adapt as needed. Reports available to the public. 8.1.a, 8.2.a-e, 8.3.a, 8.4.a-b, 8.5.a	Develop and implement a record-keeping system to provide annual progress reports to SFI and the public. Develop and implement a monitoring and feedback system to continually evaluate and mprove conformance to the standard. 19.1-2, 20.1.1-3	Both programs have annual reporting requirements, which is currently not required of the national forests.
Worker Safety, Education & Training	Provide sufficient training, guidance and supervision for all forest workers. Include safety requirements in contracts or other written agreements. Employees and contractors demonstrate a safe working environment. 7.3.a, 4.2.b	Provide sufficient education and training for employees and contractors. Develop a program for the use of certified or qualified logging professionals. 16.1.2-5	SFI has requirement that all loggers undergo safety training, which is not a current requirement for Forest Service contracts. Identified as a concern in the five pilot forests.

Comparison Category	FSC	SFI	Notes
Community Outreach, Education, & Research	Participate in local outreach, educational, economic development and civic activities. 4.1.f-g	Participate or support a variety of outreach, education, and research programs. 4.1.7, 15.1.2, 15.2.1, 16.2.1-2, 17.1.1-5, 17.2.1	
Community/Public Involvement Stakeholder Input	Seek, consider, and address stakeholder input in management planning and for site-specific management activities. For public forests, have clearly defined methods of public participation, including notice, review, comment and appeals processes. 4.4.a-d, 7.4.b	For public forests, involve the appropriate governmental entities and the public in planning and management activities. Develop a process to receive and respond to public inquiries. 18.1.1-2, 17.3.2	Covered through various National Forest Management Act, National Environmental Act, and Administrative Procedures Act requirements.
Transparency	Make summary audit, impact assessments, management plans, harvest plans, monitoring results and other documents available for public review. 6.1.d, 7.1.q, 7.4.b, 8.5.a	Provide a summary audit report to be posted on the SFI website for public review 19.1.1	Covered through various National Forest Management Act, National Environmental Act, and Administrative Procedures Act requirements.
Miscellaneous			
Statement of Commitment	Publish a statement of endorsement of and commitment to the FSC Principles. 1.6a-b	Publish a statement of commitment to the SFI 2010-2014 Standard. 16.1.1	Endorsement of the certification program by Congress, the Forest Service, or the unit would be necessary.
Indigenous People	Specific requirements for working and consulting with indigenous people. Principle 3	Communicate with affected indigenous peoples, identify and protect important sites, and address the use of NTFPs. 18.2.1	Covered through various laws and regulations related to tribal rights, uses, and consultation.
Hazardous materials storage & disposal	Storage, disposal and spill response guidelines for chemicals, containers, liquid and solid non-organic wastes including fuel and oil. 6.7a-c	n/a	The FSC has a more stringent standard about hazardous waste storage, transport, and disposal while the SFI is silent on this issue.

Table A-1. (Continued)

Comparison Category	FSC	SFI	Notes
Requirements for public lands	Must meet three threshold requirements: 1) willing landowner; 2) public consensus about management of national forests; 3) development of specific standards, indicators and guidance. Several requirements specific to public lands written in to the standard. FSC-US Federal Land Policy[c]	Several requirements specific to public lands written in to the standard.	The SFI program is ready to certify the national forests, while the FSC program would have to develop another standard prior to certifying the national forests.

Sources: http://www.sfiprogram.org/index.php and http://www.fscus.org/.

[a] FSC Pesticides Policy 2005 (FSC POL 30-001), available at http://www.fsc.org /policies.html.

[b] FSC GMO Policy 2000 (FSC POL 30-602), available at http://www.fsc.org/policies.html.

[c] FSC-US Federal Land Policy 2003, available at http://www.fscus.org/documents/.

End Notes

[1] For more information, see CRS Report R40225, *Federal Land Management Agencies: Background on Land and Resources Management*, coordinated by Ross W. Gorte.

[2] B. Cashore, F. Gale, and E. Meidinger, et al., *Confronting Sustainability: Forest Certification in Developing and Transitioning Countries* (New Haven, CT: Yale F&ES Publishing Services, 2006).

[3] See http://www.un.org/documents/ga/conf151/aconf15126-3annex3.htm.

[4] G. Auld, L. H. Gulbrandsen, and C. L. McDermott, "Certification Schemes and the Impacts on Forests and Forestry," *Annual Review of Environment and Resources*, vol. 33 (2008), pp. 187-211.

[5] See http://www.rinya.maff.go.jp/mpci/.

[6] http://www.fsc.org/facts-figures.html.

[7] C. Overdevest, "Comparing Forest Certification Schemes: the Case of Ratcheting Standards in the Forest Sector," *Socio-Economic Review*, vol. 8 (2010), pp. 47-76.

[8] C. L. McDermott, E. Noah, and B. Cashore, "Differences That 'Matter'? A Framework for Comparing Environmental Certifcation Standards and Government Policies," *Journal of Environmental Policy & Planning*, vol. 10, no. 1 (March 2008), pp. 47-70.

[9] http://www.sfiprogram.org/files/pdf/SFI_ProgressReport2011_FINAL_spreads.pdf.

[10] http://www.pefc.org/about-pefc/who-we-are/facts-a-figures.

[11] http://www.csasfmforests.ca/factsheet.htm (2010 is the most recent data available).

[12] http://www.treefarmsystem.org/aboutus.

[13] C. Overdevest, "Comparing Forest Certification Schemes: the Case of Ratcheting Standards in the Forest Sector," *Socio-Economic Review*, vol. 8 (2010), pp. 47-76.

[14] K. Fernholz et al., *Forest Certification: A Status Report*, Dovetail Partners, Inc., March 2010, http://www.dovetailinc.org.

[15] G. Auld, L. H. Gulbrandsen, and C. L. McDermott, "Certification Schemes and the Impacts on Forests and Forestry," *Annual Review of Environment and Resources*, vol. 33 (2008), pp. 187-211.

[16] Meridian Institute, *Comparative Analysis of the Forest Stewardship Council and Sustainable Forestry Initiative Certification Programs Executive Summary*, October 2001. Report is available by contacting the CRS author.

[17] Yale Program on Forest Policy and Governance, *Assessing USGBC's Policy Options for Forest Certification and the Use of Wood and Other Bio-based Materials*, February 25, 2008, http://www.yale.edu/forestcertification/ USGBCFinal.htm.

[18] C. Overdevest and M. G. Rickenbach, "Forest Certification and Institutional Governance: An Empirical Study of Forest Stewardship Council Certificate Holders in the United States," *Forest Policy and Economics*, vol. 9 (2006), pp. 93-102.

[19] A. Tikina, R. Kozak, and B. Larson, "What Factors Influence Obtaining Forest Certification in the U.S. Pacific Northwest?," *Forest Policy and Economics*, vol. 10, no. 4 (2008), pp. 240-247.

[20] G. Auld, L. H. Gulbrandsen, and C. L. McDermott, "Certification Schemes and the Impacts on Forests and Forestry," *Annual Review of Environment and Resources*, vol. 33 (2008), pp. 187-211.

[21] G. Auld, L. H. Gulbrandsen, and C. L. McDermott, "Certification Schemes and the Impacts on Forests and Forestry," *Annual Review of Environment and Resources*, vol. 33 (2008), pp. 187-211

[22] A. Roberge, L. Bouthillier, and J. Mercier, "The Gap Between Theory and Reality of Governance: the Case of Forest Certification in Quebec," *Society and Natural Resources*, vol. 24, no. 7 (2011), pp. 656-671.

[23] http://www.fsc.org/.

[24] http://www.fscus.org/.

[25] G. Auld, L. H. Gulbrandsen, and C. L. McDermott, "Certification Schemes and the Impacts on Forests and Forestry," *Annual Review of Environment and Resources*, vol. 33 (2008), pp. 187-211.

[26] B. Cashore, F. Gale, and E. Meidinger, et al., *Confronting Sustainability: Forest Certification in Developing and Transitioning Countries* (New Haven, CT: Yale F&ES Publishing Services, 2006).

[27] The geographical extent of the FSC-US standard does not include Alaska, Hawaii and the U.S. territories.

[28] http://www.fscus.org/images/documents/FSC_US_Bylaws.pdf.

[29] See FSC-US Forest Management Standard 2010 Appendix B for a map of the FSC-US regions: http://www.fscus.org/images/documents/standards/FSC-US%20Forest%20Management%20Standard%20v1.0.pdf.

[30] See http://www.sfiprogram.org/sic-maps.php.

[31] B. Cashore, F. Gale, and E.Meidinger, et al., *Confronting Sustainability: Forest Certification in Developing and Transitioning Countries* (New Haven, CT: Yale F&ES Publishing Services, 2006).

[32] See http://www.fsc.org/pcreview.html.

[33] SFI Objectives 14-20 apply to all forest operations; 1-7 apply only to forest landowners; and 8-13 apply only to fiber sourcing operations.

[34] See http://www.iso.org.

[35] See http://www.anab.org/.

[36] G.Auld, L.H. Gulbrandsen, and C.L. McDermott, "Certification Schemes and the Impacts on Forests and Forestry," *Annual Review of Environment and Resources*, vol. 33 (2008), pp. 187-211.

[37] See FSC Dispute Resolution System, FSC-STD-01-005, available online at http://www.fsc.org/accreditation_docs.html.

[38] See http://www.fscus.org/images/documents/revised_fed_land_pdfs/FL_Policy_Final.pdf.

[39] CRS was unable to verify the certification status of the DOD and DOE lands as of August 2011.

[40] R. Fletcher, M. Rickenbach, and E. Hansen, *Forest Certification in North America*, Oregon State University Extension Service, EC1518, December 2002, http://extension.oregonstate.edu/catalog/.

[41] J. Howe et al., *A Land Manager's Guide to FSC & SFI*, Dovetail Partners, Inc, September 15, 2004, http://www.dovetailinc.org.

[42] C. Overdevest, "Treadmill Politics, Information Politics, and Public Policy," *Organization & Environment*, vol. 18, no. 1 (March 2005), pp. 72-90.

[43] C. Overdevest, "Comparing Forest Certification Schemes: the Case of Ratcheting Standards in the Forest Sector," *Socio-Economic Review*, vol. 8 (2010), pp. 47-76.

[44] See http://www.yale.edu/forestcertification/USGBCFinal.htm.

[45] K. Fernholz et al., Differences Between the Forest Stewardship Council and Sustainable Forestry Initiative Certification Standards for Forest Management, Dovetail Partners, Inc., March 28, 2011, http://www.dovetailinc.org/ content/dovetail-reports-certification.

[46] C. Overdevest, "Treadmill Politics, Information Politics, and Public Policy," *Organization & Environment*, vol. 18, no. 1 (March 2005), pp. 72-90.

[47] See CRS Report R41691, *Forest Management for Resilience and Adaptation*, by Ross W. Gorte.

In: Forest Certification and Sustainable ... ISBN: 978-1- 62081-206-8
Editors: J. G. Robles and C. A. Savage

Chapter 2

FOREST MANAGEMENT FOR RESILIENCE AND ADAPTATION*

Ross W. Gorte

SUMMARY

U.S. forests are primarily temperate forests, often with relatively few species dominating over wide areas. Such forests respond and adapt to an array of environmental factors—sunlight levels and duration, temperature, precipitation, and a multitude of disturbances (e.g., fires, pests, and storms)—and these factors could be further altered by long-term shifts in natural climate variability and climate change. Many domestic forests are already under stress from drought, severe wildfires, and insect epidemics. Changing conditions and disturbances could diminish the goods and services that forests provide—timber, clean water, scenic vistas, carbon sequestration, and much more. Forests can be subject to management techniques to improve their resilience and adaptability, to assure continued production of the economically desired ecosystem goods and services.

Congress provides management direction and funding for federal forests and financial and technical assistance for management of nonfederal forests. To date, such legislative direction and funding have at most indirectly

* This is an edited, reformatted and augmented version of Congressional Research Service, Publication No. R41691, dated March 14, 2011.

encouraged management that promotes resilient forests. Congress could decide to broaden its role in promoting resilient and adaptable forests. However, thus far, no broad-based legislation has been introduced calling for direct federal forest management and nonfederal forestry assistance to sustain forests in the face of changing conditions.

Biologically diverse forests are generally more resilient (better able to recover from changes and disturbances) and more adaptable (better able to respond to changing conditions), because they have a broader biological base from which to respond. Diversity occurs at the genetic level (among trees of a given species), at the species level (among trees in a stand), and at the stand level (among the ages and sizes of trees in a stand). Many are concerned that climate change and increased climate volatility may be too rapid for natural adaptation and migration to sustain production of the desired goods and services, because some species require special habitat conditions and because forest fragmentation from human development hinders forest migration. Others argue that natural variability is sufficient to sustain forests, even in the face of climate change.

Research and monitoring are important components of understanding the extent and success of forest management efforts to promote resilient and adaptable forests. Management efforts could then respond to changing forest conditions by adjusting traditional forestry practices (e.g., prescribed burning, thinning, timber harvesting, tree planting, and more) or even by taking more intensive action to assist forest adaptation (e.g., reducing habitat fragmentation, creating habitat corridors, and assisting species relocation). A question for Congress and resource managers is whether or how to fund and implement research and monitoring programs to provide information for forest management in a time of changing conditions.

Forests respond to various factors, such as amount of sunlight, temperature levels and variations, precipitation amounts and timing, and a host of disturbances (fire, pests, Finvasive species). Changes in climatic patterns—whether natural or human-induced—could affect these factors, possibly increasing temperatures, altering the hydrologic cycle, and exacerbating forest disturbances. Such changes could alter the dynamic balance of forest ecosystems, shifting forest habitats north and to higher elevations. Many are concerned that the natural ability of U.S. temperate forests to adapt to changing climate conditions may be insufficient to sustain production of desirable ecosystem services—timber, recreation, water, carbon sequestration, and more. Some predictions show climate-induced habitat shifts outpacing the natural rate of forest migration and dispersal, making forests vulnerable to pests, diseases, and invasive species and to extreme events, like wildfires, all of which can damage ecosystem services, such as water quality and desired animal habitats.[1] Others contend that ecosystems have been able to adapt in the past, and they will be able to adapt again.

Forest management can affect the ability of forests to respond and adapt to changing conditions. To date, legislative direction and funding for management of federal lands and for federal assistance in managing nonfederal lands has, at most, indirectly encouraged management for resilience (recovery from) and adaptation to changing conditions. Legislation on forest management generally has been either broadly focused—very general direction in climate or energy bills, for example—or specific to regional or local forestry problems, such as the extensive mountain pine beetle infestation in the northern and central Rocky Mountains. This report focuses on management approaches that might be employed to promote forests that are resilient and adaptable to disturbances and changing conditions, regardless of the cause.

Forests can be managed to improve their resilience and adaptation to disturbances and possible climatic change and variability. Diverse forests can sustain the production of desired ecosystem services, such as clean water, wildlife habitat, recreational opportunities, timber and biomass, carbon sequestration, and more. Forest management for diversity could include changes in wildfire management, forest health, timber production, and forest regeneration; more intensive adaptation practices include reducing forest fragmentation, providing habitat corridors, and assisting migration. These practices are discussed in Appendix B.

FORESTS OF THE UNITED STATES

Domestic U.S. forests are dominated by relatively few forest types (compared to tropical forests), but contain an array of species mixes, as described in Appendix A. Within each of the many temperate forest types, often relatively few tree species dominate the overstory, and more tree and shrub species occur in the understory. The species and species mixes have evolved and adapted to survive within certain climatic conditions and complex ecological relationships. Historic temperature patterns and forest dynamics suggest that habitats are changing and will continue to change. However, some have expressed concerns that the predicted rate of change will exceed the natural adaptive rate, decreasing total temperate forest area.[2] Thus, changes in climatic patterns are seen as potentially altering domestic forests and possibly depreciating their value in providing economically valuable goods and services.[3]

The 750 million acres of U.S. forests are owned and managed by many entities—the federal government, states, local governments, corporations, and other private landowners. The relative dominance of these landowner groups varies regionally, as shown in Appendix A. Federal forests are most common in the West. States and local governments own significant forestlands, especially in the Midwest (Minnesota, Michigan, and Wisconsin) and mid-Atlantic (Pennsylvania and New York) regions. Corporate forests tend to be concentrated in the most timber-growing productive areas—the Pacific Coast and the South—and are generally managed for industrial wood products. Other private forests dominate in the eastern regions. These landholdings vary widely in size and purpose, from small holdings (e.g., summer homes) that may occasionally sell timber for extra revenue to tens of thousands of acres managed by hunt clubs, land trusts, and other organizations.

The principal federal forest management agency—the Forest Service in the Department of Agriculture (USDA)—is attempting to address the possible effects of climate change on land and resource management.[4] In July 2010, the Forest Service issued its *National Roadmap for Responding to Climate Change.*[5] This strategic plan provides general direction for national forest adaptation, for agency capacity-building, and for cooperation among landowners since climate changes affect all forests.

The management intensity and expertise applied to private forests vary widely. The Forest Service and Natural Resources Conservation Service (also in USDA) have numerous programs to provide technical and some financial assistance to non-corporate private forest owners, directly and through state forestry organizations. Assistance programs could provide support for forest resilience and adaptation by non-corporate forest owners.

Impacts of Climate Variability and Change on Temperate Forests

Domestic forests have experienced significant changes in the past 30 years or more, with greater tree mortality from drought, wildfires, and a host of insects and diseases.[6] Even under modest greenhouse gas (GHG) emissions scenarios, climate change is projected to increase temperatures and climate volatility, alter hydrology and water availability, and substantially expand the extent and impacts of disturbances.[7] Change is not inherently bad, and can in many instances lead to desirable new forests and ecosystem services. The new forests would likely be a mix of existing natural forest species and exotic species, and the result could be radically different ecosystems. It is unclear whether these new ecosystems would continue to provide timber, wildlife habitat, and other economically beneficial values.

Possible Temperature Changes

There is a broad scientific consensus that average temperatures in the lower atmosphere have increased over the last century, and this trend is expected to continue and possibly increase.[8] Forest species are adapted to specific habitats, generally defined by a particular pattern of temperature, hydrology, and other climatic factors. Warming temperatures could both alter existing habitats and their corresponding species mixes and spur habitats and species to migrate. In general, rising temperatures are expected to cause habitats to migrate upslope and expand northward, while contracting at their southern extremes.[9] Generally, oak-hickory, oak-pine, ponderosa pine, and arid woodland habitats are expected to expand, while alpine and subalpine habitats, like spruce-fir and aspen-birch, are likely to decrease.[10] One study has estimated that, by the end of the century, oak-hickory forests in southern New England could move 70-150 miles north, substantially increasing these forests and decreasing the area of other forest types (e.g., maple-beech-birch and spruce-fir forests).[11]

Research on forests of the northeastern United States provides an example of what has already happened and what more might happen. In Vermont's Green Mountains, temperatures rose by 1.1° C between 1964 and 2004, leading the spruce-fir and maple-beech-birch forest types to move upslope by more than 300 feet in elevation over the 40-year period.[12] Even larger temperature increases are projected to occur for the region by 2100.[13] Such temperature increases could mean that, by the end of the century, oak-hickory forests in southern New England could move north and increase their area by 50% to 400%.[14] This would likely reduce the area of maple-beech-birch forests (and maple syrup production) by a quarter and move them upslope by more than 250 feet in elevation. This would coincide with retreat of nearly 2000 feet higher in elevation for boreal conifer (spruce-fir and white-jack-red pine) forests, and thus their elimination from many mountaintops. A similar retreat in alpine-tundra habitats has been reported in the Sierra Nevada Mountains of California.[15] Conversely, a downslope expansion of several species in California reportedly occurred despite warmer temperatures, because of an accompanying shift in hydrologic patterns (more rain and snow at lower elevations).[16]

Possible Hydrologic Changes

Experts predict that temperature changes will probably be accompanied by hydrologic changes, adding more stress to trees in some areas, while benefitting trees in other areas. Warming trends are expected to increase the volatility of precipitation events and could lead to both more rain and more drought.[17] There is less agreement on precipitation projections among climate change models than there is on temperature predictions.[18] For example, for the Northeast, precipitation predictions range from roughly 25% increases by 2100, to little change or small regional decreases.[19]

The severity and frequency of drought is expected to increase, especially in areas where drought is already an issue. This trend will likely add stress to U.S. forests, reducing tree growth and increasing tree mortality. "The primary immediate response of trees to drought is to reduce net primary production [plant growth].... Under extended severe drought conditions, plants die."[20] Cases of widespread tree mortality have often occurred during the most severe droughts. Two examples are the extensive tree mortality in the southern Appalachian Mountains and the Great Plains during the dust-bowl droughts of 1920s and 1930s, and the die-off of multiple pine species in the Southwest during the 1950s drought.[21] Climate models also project more frequent and intense ice storms, wind storms, and heavy rainfall, [22] all of which can increase disturbances and tree mortality.

Furthermore, increasing temperatures are expected to shorten U.S. winters, decreasing snowpack and hastening snowmelt, albeit with substantial regional variation. Early snowmelt and reduced snowpack are expected to increase summer moisture deficits (seasonal droughts), increasing stress on forests, increasing tree mortality, and making forests more vulnerable to disturbances.[23] Early snowmelt also has been shown to increase the danger of landslides, flash floods, and wildfire.

Forest Disturbances

Disturbances are events that kill anywhere from a few trees to extensive stands, and include sudden events (e.g., severe storms and floods) as well as more drawn-out occurrences (e.g., wildfires, insect or disease infestations, and drought). Disturbances of various sorts occur naturally in temperate ecosystems. Disturbances kill and knock down trees, providing seedlings access to sunlight and space; when they occur at a modest scale, disturbances allow forests to adapt far more rapidly than without the disturbance.[24]

Pests—insects and diseases—are U.S. forests' greatest natural disturbance element. Insects and diseases, both native and exotic, can limit tree growth and kill trees. In 2008, nearly 9 million acres of trees (more than 1% of all U.S. forests) were killed by insects and diseases, a 32% increase from 2007.[25] Unusually hot, dry weather patterns over a few years can increase insect survival rates, and are believed to be responsible for increased insect outbreaks in forests in the U.S. Southwest and West.[26] Current data and models suggest that global warming will result in the redistribution of insect pests, resulting in the invasion of new habitats and forest types.[27] The mountain pine beetle is an example of how climate variability and change can convert a regionally native insect into an invasive species.[28] For the beetles, long, cold winters limit overwinter survival, and therefore have geographically restricted the insect.[29] Warmer temperatures accelerate reproductive success, increasing the infestation while reducing the trees' ability to defend themselves.[30] The current infestation has reached parts of northern Canada previously unaffected, and possibly allowing the beetle to invade eastern Canada and the eastern United States, where it has never existed before.

Wildfires are another significant natural forest disturbance. Climate variability has historically been a key factor for wildfires in the western United States.[31] Higher temperatures and drier conditions are expected to bring more severe fire weather.[32] In addition, the variability between fire seasons is expected to increase.[33] Warmer springs and earlier snowmelt have been shown to provide a productive growing season (i.e., a wet spring) often followed by a seasonal drought, turning the large quantities of biomass (plant matter) from the growing season into fuel for a potential forest fire.[34]

Some forests regenerate after extensive disturbances with fewer species, usually the strongest survivors, in a process called synchronization. A synchronized forest generally has a relative dearth of species and only one or a few age classes. Increased climate volatility, with more frequent and more extreme events to disturb forests,[35] could increase the synchrony of U.S. forests, increasing their vulnerability by reducing the number and variety of a forest's defenses. For example, mountain pine beetles attack lodgepole pines of more than 5 inches in diameter; the synchronized regeneration of lodgepole pine forests following severe fires a century ago is a major factor in the current mountain pine beetle epidemic and extensive forest mortality.[36]

Atmospheric Changes

Greenhouse gas emissions can also affect forests. Increased CO_2 and nitrogen levels are expected to increase average growth rates (net primary productivity) in U.S. forests.[37] This would likely increase carbon sequestration by U.S. forests, mitigating or offsetting some greenhouse gas emissions.[38] Furthermore, increased atmospheric concentrations of CO_2 can also increase plant water-use efficiency.[39] This could mitigate stress from any declines in precipitation, but higher temperatures and longer growing seasons also increase total water demands for sustaining plant growth.[40] In addition, higher ozone concentrations can reduce growth rates, while nitrogen deposition can acidify soils, reducing forest productivity and degrading water quality.[41]

Because of the possible counteracting effects of increasing carbon, nitrogen, and ozone levels in the atmosphere and of increasing atmospheric carbon and temperature, it is unclear whether atmospheric changes would benefit or harm forests. One report noted that "increases in biomass across many forest types … [are] attributed to climate change. However, without knowing the disturbance history of a forest, growth could also be caused by normal recovery from unknown disturbances."[42]

FOREST RESILIENCE AND ADAPTATION

The Importance of Biodiversity

Forest resilience—the ability to respond to and recover from disturbances and changing conditions—and forest adaptation—the ability to change or migrate in response to disturbances and changing conditions—historically have been key to ecosystem sustainability.[43] Conversely, unsuccessful adaptation may lead to extinction.[44] Diversity among and within plant and animal species in an area is an important aspect of providing for resilience and adaptation. Biodiversity has been described as "the fundamental building block of the services that ecosystems deliver."[45] In contrast, fewer species tend to concentrate or exacerbate risks.[46] Higher levels of biodiversity generally contribute to ecosystem health and stability by providing a broader base, or toolbox, from which to respond, recover, or adapt. In other words, diverse ecosystems are more stable (resilient) over time and more adaptable (better able to evolve) in response to changing conditions, because they contain more variations that can potentially survive.[47]

Biodiversity occurs at three primary levels. The lowest or most basic level is *genetic diversity*—the variations in genetic composition among trees of the same species. Because of this variability, some individuals are better adapted to certain conditions—more tolerant to drought, more resistant to pests, and so forth. For example, individual white pine trees vary significantly in their resistance to white pine blister rust.[48] Trees from the seeds of resistant individuals are also resistant, and planting resistant trees is a common approach to reducing pest problems. This approach is being tried to reestablish the American chestnut, which was virtually eliminated from eastern U.S. forests by an introduced fungus (the chestnut blight) early in the 20th century.[49] Clones and genetic monocultures are an extreme form, with no genetic diversity whatsoever (each plant is genetically identical to every other plant). Such practices have been used in agriculture for decades, and the condition occurs naturally in some plants (e.g., many aspen stands are multiple trees from a common rootstock, and thus are effectively monocultures); however, the lack of genetic variation in crops and livestock has also long been recognized as increasing risks from pests.[50]

The second level of biodiversity is *species diversity*—the variations in species within a stand of trees. Species diversity in temperate forests is modest, compared to the diversity of tropical forests. Nonetheless, as described in Appendix A, most U.S. temperate forest types include several major species. Even those types with a single dominant species typically contain several additional species. Species diversity tends to protect forests against pest epidemics, as most pests are host-specific, attacking only one or a few closely related species. Species diversity promotes the adaptability of forests to pests, fire, drought, and other disturbances, since different species respond differently to the disturbances.

The third or highest level of biodiversity is *stand diversity*—the variation in tree ages and sizes among stands of trees, as well as in species. As with genetic and species diversity, stand diversity (also known as asynchrony) promotes forest resilience by providing a variety of conditions to respond to changes and disturbances. For example, the mountain pine beetle (discussed above) only infests lodgepole pine trees greater than 5 inches in diameter. Thus, a lodgepole pine forest containing stands of varying sizes can withstand an epidemic, because some stands will not become infested.

Possible Congressional Role

Congress could have an expanded role in promoting resilient and adaptable forests. Congress establishes direction and provides funding for federal forests and for financial and technical assistance for management of nonfederal forests. For federal forests, current congressional direction indirectly encourages management that promotes resilient forests. For the USDA Forest Service (59% of federal forests), the basic direction is in the Multiple Use-Sustained Yield Act of 1960 (P.L. 86-517; 16 U.S.C. §§528-531): to produce a sustained yield of the multiple uses (goods and services) from the lands, without impairing long-term productivity. While not explicit direction to manage the national forests to assure resilience and diversity, the direction for sustained yields implies management for adaptation to changing conditions, including a changing climate. However, this and many other laws govern Forest Service management activities, and some observe that the total legislative direction is at best cumbersome and at times even contradictory.[51]

Congress also authorizes and funds technical and financial assistance for managing nonfederal forests.[52] However, the federal government has no direct role in management of nonfederal forests; regulation of forestry practices on private lands is exclusively a state prerogative and responsibility.[53] Nonetheless, provisions enacted in the 2008 farm bill (P.L. 110-246, the Food, Conservation, and Energy Act of 2008) require states to develop assessments of forest conditions, trends, threats, and priorities to receive funding.[54] The national priorities for assistance include managing forests for multiple goods and values; protecting forests from disturbances; and enhancing production of public benefits. Climate change was listed explicitly as an issue for forestry assistance—clearly direction for federal assistance for nonfederal forest management to provide adaptable, resilient forests.

Legislation in the 112th Congress could affect management for forest adaptation, though to date no bills have been introduced with broad forest management implications. None of the bills introduced in the 111th Congress provided guidance for federal management or assistance for forest adaptation to climate change. At a broad scale, the climate cap-and-trade bills in the 111th Congress, such as S. 1733 and H.R. 2454, generally included funding for forest adaptation, but with very little guidance on what to do or how to use the funding.[55] At the other end of the scale, bills were introduced to address specific regional or local forest health problems, such as the mountain pine beetle epidemic in the northern and central Rocky Mountains (see, for example, H.R. 5192, S. 2724, and S. 2798).

Limits to Natural Adaptation

While forest ecosystems can adapt to changing climatic conditions over the long run (a century or more), the desired goods and services provided by forests may well be disrupted as forest ecosystems adapt. Some are concerned that the expected rate of climate change may be too abrupt for U.S. temperate forests to adapt or migrate without substantial losses.[56] Tree species dispersal is very slow, due to long generation intervals, and thus conditions may change more quickly than tree species can disperse. Most trees alive today will still be alive in 2090, when global climate models predict significantly higher temperatures and altered hydrological patterns.[57]

Species and populations most likely to be harmed by climate change include habitat specialists that require special conditions; populations of trees on the southern and/or southeastern edges of their habitat, such as the boreal forests of the northeastern United States; species located in fragmented habitat (see "Fragmentation and Corridors," below); and species that are poor colonizers or dispersers (i.e., that have difficulty regenerating).[58] Ecological community interactions play an important role in determining ecosystem migration. Such community interactions as species competition, nutrient chain relationships, and symbiotic relationships may limit migration.[59]

Forest management could assist resilience and adaptation to disturbances and to the accelerated rate of change projected by most climate models. Diverse, sustainable forests can continue to produce the desired array of ecosystem services, such as clean water, wildlife habitat, recreational opportunities, timber and biomass, carbon sequestration, and more.

Management for sustainable forests suggests that "ecological adaptation" might replace "ecological restoration" as the underlying management approach to conservation and sustained production. For many years, various interests have approached forest management under the presumption that restoring forest ecosystems to historical conditions, commonly defined as pre-Euro-American settlement, is the best way to promote and sustain ecological integrity and forest productivity.[60] However, climate change might substantially alter regional climates from the conditions of the last several centuries, or even millennia.[61] Thus, "restoration" of historic conditions could be unsustainable or infeasible. Instead, forest management approaches could emphasize responses to disturbances to provide and enhance forest resilience and diversity at the genetic, species, and stand levels.

Monitoring and Responding to Results

The potential effects of climate variability and change on forested ecosystems are broad and varied. Uncertainty over changes in temperature, levels and timing of precipitation, and catastrophic events, combined with incomplete knowledge of ecological systems, leads to even greater uncertainty about natural ecological responses. Efforts to manage uncertainty include several continuing activities: conducting research on ecological conditions and changes; monitoring the results of climate change and of management efforts; and adjusting management based on the information gathered.

Research and Information

Reducing threats, increasing diversity, and improving habitat connectivity are based on relatively simple, broadly accepted understandings of species, ecosystems, and the possible effects of climate change.[62] However, there is substantial uncertainty as to the ability of forests to recover from and/or adapt to changes, due at least in part to the uncertainty of climate change impacts themselves. Annual variability and the unknowns about the long-term nature of climate change suggest an approach of monitoring changes in conditions and results, and of adjusting management to assist forests in recovering from and adapting to changing conditions. However, such research and monitoring can also be costly.

The research and information needed to manage for forest resilience and adaptation cover broad subjects. Four topic areas have been identified by researchers as significant research gaps:

1. *Baseline Data.* Baseline data on the current distribution of species and species mixes do not exist for many U.S. forests. Some argue that baseline data are necessary to measure the effects of changes in climate, to inform management strategies, and to improve predictions about the effects of changes.[63]
2. *Climate Change Predictions.* Reduced uncertainty in climate change predictions could facilitate decision making. Generally, uncertainty in predicting the magnitude and extent of climate change restrains action. Some scientists maintain that reducing uncertainty in predictions of temperature and precipitation could make forest adaptation management less risky and less uncertain.[64]
3. *Forest Ecosystem Responses.* Management strategies could be informed by further research on forest ecosystem responses to

stressors, such as chronic or acute water stress, belowground processes, habitat migration, pests and diseases, and growth rate responses to atmospheric changes.[65] Historic and paleo-ecological data can document past forest responses to different climatic conditions, and thus may be useful for predictions about future ecological changes.[66]

4. *Adaptive Management Efficacy.* Relatively little is known about the effectiveness of adaptive management.[67] It is a relatively modern development in forestry, and research on the effects of forest management can take several years to decades, due to the long generation cycles of forests.

Monitoring Conditions and Results

A lack of existing systematic measurement and monitoring programs may constrain the ability of scientists and managers to determine if forests are adapting to climatic and other changes, either naturally or as a result of management strategies. On-the-ground monitoring can provide insight into the stresses experienced by trees and the nature and direction of ecosystem change, thus providing information to guide management responses.[68]

Several monitoring programs exist, some of which might be a useful beginning for examining climate variability and impacts on forests, and the effectiveness of natural and managed adaptation, but none are specifically designed for this purpose.[69]

- The USDA Forest Service's Forest Inventory and Analysis Program (FIA) began in 1917. FIA reports on the status and trends of tree species size and volume. For all classes of forest land ownership, FIA measures and reports total tree inventory, growth, mortality, and removals by harvest, as well as wood production and utilization rates by various products.[70] FIA is comprised of a series of statistically validated permanent forest plots. It is widely considered to be successful at accurately measuring timber data, which is what it was designed to do. While valuable for some purposes, its has limited utility for identifying baseline forest conditions and measuring other changes because of its historic emphasis on commercial timber volumes.
- The European Union's intensive forest health monitoring network, EU/ICP-Forests Level II, is another ecosystem monitoring example. This program monitors habitat shifts for a wide spectrum of species

across the European Union (about 1.07 billion acres) with 6,000 monitoring plots.[71] It has also provided data and scientific rationale for policy decisions, promoted public awareness of ecosystem health, and created an early warning system for environmental issues, such as forest destruction, biodiversity changes, and deposition (precipitation of pollution).[72]

- The Northwest Forest Plan was created by a record of decision for the Bureau of Land Management and the Forest Service by the Clinton Administration in 1994. This adaptive management plan instructed forest managers to experiment with, monitor, and interpret forest management strategies over an area of 10 million acres of federal forests in the Pacific Northwest.[73] The plan established a large-scale regional program to monitor old-growth forests, northern spotted owl and marbled murrelet habitats, watersheds, and socioeconomic effects on local communities, tribes, and others.[74] After 10 years, an assessment of the plan criticized managers for exercising too much caution in adaptive management strategies and experimentation, but found the established large-scale monitoring system to be extremely useful in both expected and unexpected ways.[75] For example, a balance sheet resulting from the plan showed that forest growth outpaced losses to timber harvest, fire and other disturbances, something that many foresters had expected but that surprised others.

Other landscape-scale programs include monitoring that could prove useful to assessing ecological responses to management and environmental changes; examples include the Sierra Nevada Forest Plan and the Southern Appalachian Forest assessment. In addition, other USFS programs provide information on forest conditions and responses to stimuli. For example, the National Fire Plan and the Forest Health Monitoring Program (focused on insects, diseases, and invasive species) could provide consistent approaches for assessing changing conditions and the results of management. Nonetheless, these efforts are geographically or topically limited, and may be of limited value for determining baselines and changing conditions.

Management Response

Forest management, whether plantations for timber production or protected natural areas, affects species diversity and stand structures. Traditional forestry practices can be modified to emphasize diversity for resilient and adaptable forests. For example, wildfire management can be modified to include prescribed burning and *appropriate management response* (see Appendix B) to increase natural fire benefits. Similarly, timber harvesting, thinning, and stand regeneration practices can emphasize species for adaptable responses to disturbances and changing conditions, while forest certification is a means of providing oversight to assure sustained forest productivity. Some advocate more intensive actions to support forest adaptation, such as reducing forest fragmentation, protecting habitat corridors, and even managing species relocation. These possibilities are discussed in more depth in Appendix B.

Many argue that forest management strategies need to respond and adjust to new insights about forest health from research and monitoring.[76] The uncertainties of projecting future climate change and its effects on ecological systems suggest flexibility in management plans. Forest adaptation likely could be improved with up-to-date climate science and information about local changes for land managers. Some suggest that managers should be involved in monitoring changes in the forests themselves, because they see and know what changes are occurring. [77] To do so, forest monitoring efforts would need to be expanded and the focus altered. However, such monitoring might be costly, and could divert managers' attention from decision making and implementation.

"Adaptive management" is a commonly described management style of taking action, assessing the results, and modifying the future actions based on the results.[78] Passive adaptive management implements the "best apparent management option," with feedback from research and monitoring to adjust future management activities.[79] Active adaptive management allows for experimentation with monitoring, and allows management strategies to be altered quickly in response to new information—both research and on-site results—to address expected and ongoing changes. Forest management may need to be adaptive. Although important lessons can be learned from the past, past conditions might not be replicable in a changing future.

APPENDIX A. DOMESTIC U.S. FORESTS

Most domestic U.S. forests are temperate forests, with relatively few tree species (compared to tropical forests), but various species mixes. The text box below describes the temperate forests that dominate most of the United States. Alaska contains mostly boreal forests of hemlock-Sitka spruce along the southern coast and panhandle, and of spruce-birch mixes in the interior. The

United States also includes some tropical forests in southern Florida and Hawaii.

Within each of the many temperate forest types, a few to several tree species dominate the overstory, and several to many additional plant species occur in the understory. These species and species mixes have evolved and adapted to habitats with certain climatic conditions and complex ecological relationships. Historical evidence of temperature patterns and forest dynamics suggest that forest habitats have changed and are continuing to change. Some observers, however, are concerned that the rate of climate change exceeds the natural adaptive rate, and will cause a decrease in total temperate forest habitat.[80] If so, climate change could substantially change domestic forests and possibly degrade their value in providing desired goods and services.

Temperate Forest Types in the United States

Several forest types dominate each of the ecological regions of the United States. The Northeast and North Central regions contain 172 million acres of forest, dominated by oak-hickory forests (32% of the area) in the southern portions of the regions and maple-beech-birch forests (31%) in the northern portions. Other important northern forest types include aspen-birch forests (10%), spruce-fir forests (9%), elm-ash-cottonwood forests (7%), and white-jack-red pine forests (6%). In the Southeast and South Central regions, the 215 million acres of forest are dominated by oak-hickory forests (39%) and loblolly-shortleaf pine forests (26%), but also include oak-pine forests (11%), oak-gum-cypress forests (10%), and longleaf-slash pine forests (6%). The highly productive 87 million acres of forest in the Pacific Coast states are more variable; they are dominated by Douglas-fir forests (24%) in the northern portions and western hardwood forests (23%) in the southern portions, but also include ponderosa pine forests (11%), fir-spruce forests (9%), piny on-juniper forests (7%), hemlock-Sitka spruce forests (6%), and more. The 145 million acres of forest in the Intermountain region are also highly variable—dominated by pinyon-juniper forests (33%), but including fir-spruce forests (15%), Douglas-fir forests (19%), western hardwood forests (18%), ponderosa pine forests (11%), and lodgepole pine forests (8%).

Source: W.B. Smith et al., *Forest Resources of the United States, 2007: A Technical Document Supporting the Forest Service 2010 RPA Assessment* (Washington: USDA Forest Service, 2009), Tables 5 and 6, pp. 163-164, 171.

U.S forests are valuable assets. Forests provide many goods and services (collectively known as "ecosystem services"), such as timber, clean water, recreation, and more. The ecosystem services provided by forests may be vulnerable, if forests cannot adapt to the changing climate. However, management to ameliorate disturbances and to increase forest diversity and resilience could help avoid or minimize such possible losses.

Most climate scientists and forestry experts agree that climate change could have significant ecological consequences, altering species mixes and increasing forest disturbances. While U.S. temperate forests have existed when the earth had significantly higher CO_2 concentrations and was significantly warmer, and will undoubtedly exist in a different climatic future, the issue is the nature and rate of the transitions from current forest conditions to future conditions. Forests take decades to adapt, and thus could be in transition for the next century or longer.[81] The important question is: will U.S. forests continue to provide economically valuable ecosystem services as they adapt to climate and other changes?

Although changing climatic conditions would likely affect all regions, the impacts are likely to vary by region and within regions, due to differences in latitude, hydrology, and topography, as well as in the impacts on individual species and species mixes. Particular species could be affected by specific disturbances, significantly altering the ecosystems. One historic example of the impact of an invasive species is the chestnut blight, which virtually eliminated chestnuts that had accounted for as much as a quarter of the trees in oak-hickory forests a century ago. As discussed further below, climate change makes the introduction of invasive species more likely. Different forest types are likely to be affected differently by climate change and may require different management strategies to promote their adaptation.

Domestic Forest Ownership

U.S. forests are owned and managed by many entities, including the federal government, states, local governments, corporations, and other private landowners. The relative importance of these landowner groups varies regionally, as shown in Figure A-1. The largest federal forest management agency is the Forest Service, in the Department of Agriculture (USDA). These forests are primarily in the West, although the Forest Service is the largest landowner in the eastern regions. The Department of the Interior (DOI), particularly the Bureau of Land Management (BLM), administers most of the other federal forestlands, although the Departments of Defense and Energy also manage federal forests. Several states and local governments also own significant amounts of forestland; this is especially the case in the Midwest (Minnesota, Michigan, and Wisconsin) and Mid-Atlantic (Pennsylvania and New York) regions.

Private forests are more common in the North and South regions. Corporate forests are substantially concentrated in the most productive areas for timber growing—the Pacific Coast (Washington, Oregon, and California) and the South—and are generally managed for industrial wood products. Other private forests—also called non-industrial private forests—dominate in the eastern regions. These landholdings vary widely in size and purpose, ranging from a few acres for a second home on a forested lakeshore or a woodlot on a farm, which occasionally provide timber revenues for the landowners, to tens of thousands of acres managed by hunt clubs, land trusts, and other organizations, where timber production is often a secondary or residual result of management.

Forest Management Programs

The Forest Service and the Department of the Interior are addressing the effects of climate change on federal land and resource management.[82] The Forest Service released its *National Roadmap for Responding to Climate Change* in July 2010.[83] This is a strategic plan that establishes general management guidance for national forest adaptation and for carbon mitigation from agency activities. It also acknowledges the need to build agency capacity and seeks to build cooperation among landowners recognizing that climate change affects all forests. The Interior Department established a Climate Change Task Force, which developed three reports analyzing the impacts of climate change and identifying relevant options for DOI agencies.[84] The department has also created a climate change response council, eight regional climate science centers, and a network of landscape conservation cooperatives for managing climate-change impacts.[85]

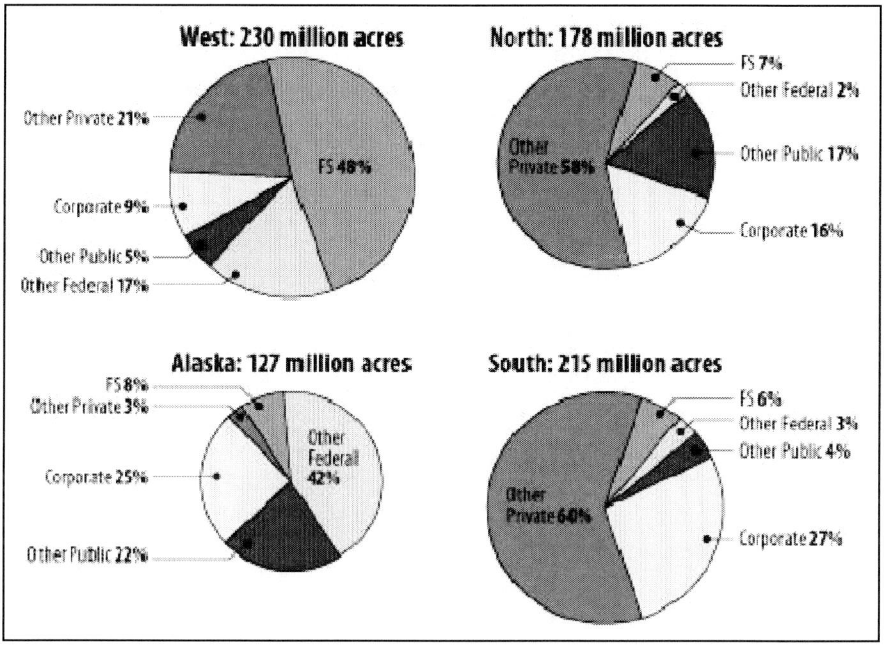

Source: W.B. Smith et al., *Forest Resources of the United States, 2007*, USDA Forest Service, Gen. Tech. Rept. WO-78, Washington, DC, 2009, pp. 154-156, http://nrs.fs.fed.us/pubs/gtr/gtr_wo78.pdf.

Figure A-1. U.S. Forest Ownership by Region.

The management of private lands varies widely. Some, such as many corporate timberlands, are managed intensively and rely on substantial forestry expertise. Others, including many of the non-industrial forests, are largely left to grow with little management effort. The Forest Service and Natural Resources Conservation Service (also in USDA) have numerous conservation programs that can provide technical and some financial assistance to private forest owners. Some assistance is available directly from the federal agencies, while some is provided through state forestry agencies. To the extent that these agencies recognize and are attempting to respond to climate variability and change, federal assistance programs to private forest owners are likely to provide information and support for forest adaptation.

APPENDIX B. MANAGEMENT FOR FOREST ADAPTATION

Diversity of species and stand structures allow forests to be resilient and adaptable, whether intensively managed plantations for timber or protected natural areas. Traditional forestry practices, such as wildfire management and timber harvesting, can be tweaked to emphasize structures and diversity that enhance the ability of forests to respond and adapt to change. Landowners undertake many activities in response to disturbances, especially wildfires and insect and disease epidemics. Wildfire and forest health management practices to control damages and prevent or ameliorate possible damages affect biodiversity, and thus can be used to increase the resilience and adaptability of forests. Timber management can enhance or reduce forest adaptability, depending on the practices used and how they are implemented. More intensive actions, such as providing migration corridors or even assisting forest migration, may be warranted in some circumstances. Finally, monitoring is an important aspect to determine how forests are responding to management activities and changing conditions, to assure the sustained production of desired forest ecosystem services.

Wildfire Management

Wildfires are natural disturbances in nearly all temperate forests, with wide variations in intensity and frequency. For much of the 20th century, fire was seen as inherently destructive to forests, and efforts were made to control and prevent all wildfires. This history of aggressive fire suppression, combined with historic logging and grazing practices, have made some ecosystems susceptible to unnaturally intense wildfires.[86] Alternative wildfire management approaches—altering fire protection and control and reducing fuel levels—can enhance the resilience and adaptability of some forest ecosystems.[87]

Fire Protection and Control

Wildfires do not burn uniformly; when burning within natural parameters (which may be difficult to define), wildfires tend to increase species and stand diversity and thus enhance forest resilience and adaptability.[88] Thus, some have advocated that fires which do not threaten lives and resources be allowed to burn, saving taxpayer dollars and providing the biodiversity that comes from natural wildfires.[89] In addition, because of weather conditions, fuel loads, and in some ecosystems, natural vegetation patterns, many crown fires cannot be controlled, although sometimes such fires can be guided away from high value areas and structures.

Federal land management agencies have developed an approach that allows some wildfires to burn—the *appropriate management response (AMR).* The approach "encompasses all the response actions necessary to manage a wildfire … from monitoring a fire at a distance to intensive suppression actions."[90] AMR, at least in theory, allows fire managers to aggressively suppress fires, or portions of fires, that threaten lives, property, and resources; to use fire control methods to direct fires away from threatening situations; and to use fires, or portions of fires, to achieve management objectives, such as reducing fuel loads, when lives, property, and resources are not threatened. Critics of federal fire management, however, suggest that implementation of AMR has been rather limited.[91]

Fuel Reduction Efforts

Some ecosystems have unnaturally high biomass fuel levels, because of aggressive suppression of all wildfires and past logging and grazing practices. In some ecosystems, reducing fuel levels to lower the severity of wildfires to historically natural intensities can generate the benefits of stand asynchrony and heterogeneity associated with natural wildfires.[92] Fuel reduction efforts have been the focus on much of the debate over increasingly severe wildfires in recent years.[93] The two primary tools for reducing fuels are prescribed burning and mechanical thinning (described below under "Timber Management").

Prescribed burning is deliberately setting fires in specific areas under identified conditions. (Some include wildfires not actively suppressed under Appropriate Management Response, described above, as prescribed burning, but the federal agencies use the term "wildland fire use" for such situations.) Prescribed burning is widely used for fuel management, because it reduces biomass (the fuels) to ashes (minerals), which act as fertilizer. It is particularly effective at reducing the smaller fuels, especially in the arid West where deterioration by decomposers (insects, fungi, etc.) is often very slow, and may well get slower if changing climatic conditions exacerbate drought. In fact, prescribed burning is the only treatment that directly reduces the fine and small fuels that are known to facilitate wildfires. Prescribed burns may also reduce the risk and extent of disturbances like wildfires and infestations of pests, diseases, or invasive species.[94] Prescribed burns alter forest structure by removing some of the understory in a spatially heterogeneous manner, increasing diversity for more resilient, adaptable forests.

Prescribed burns are not perfect tools. They are not fully controlled actions, and occasionally escape and cause extensive damages; for example, the Cerro Grande fire that burned 237 houses in Los Alamos, NM, in May 2000 was an escaped prescribed burn. Also, some are concerned that the emissions from prescribed burns may exacerbate climate change. Both wildfires and prescribed burns release large quantities of CO_2. Advocates of prescribed burning contend that the burns merely shift the timing of fire, not the total amount burned—that the CO_2 released by prescribed burning equals what would have been released by wildfires without prescribed burns. Further, many argue that prescribed burns significantly reduce total CO_2 emissions by reducing the frequency of severe fires.[95]

Forest Health Management

Forest health management is the term used for preventing, controlling, or eradicating insects, diseases, and invasive species. In contrast to wildfire management, the tools for forest health management are highly variable and often specific to a particular pest. For example, insects might be defoliators (destroying leaves or needles) or borers (infiltrating under the bark); diseases might defoliate or damage tree roots. Many pests are host-specific, attacking only one species of tree, while others can infest a wider array of hosts; for example, Douglas-fir tussock moth only feeds on Douglas-fir trees, while white pine blister rust can infest all white pine species, and gypsy moths can feed on nearly any tree species.[96] Invasive species might be insects or diseases, or they could be plants or animals that displace or feed on native species.

The tools for forest health management generally include manual controls, chemical controls, and biological controls. Manual controls include cutting and removing infested trees (often called salvage and sanitation harvesting, and discussed below) or invasive plants. Chemical controls include pesticides (insecticides, herbicides, etc.) of various sorts; special care is commonly taken with chemical use, because of the potential for effects on human health.[97] Biological controls include an array of species to introduce pathogens, parasites, or predators to eliminate the target species. These approaches have varying degrees of specificity in controlling pests—manual methods can be very specific, but many chemicals are broad-spectrum. Biological controls can be very host-specific (e.g., parasitic wasps that only live on a particular species) or more broadly controlling (e.g., Bt [Bacillus thuringiensis] infects all Lepidoptera [moths and butterflies]).[98] A longer-term strategy, and the only effective one for some diseases, is the selection and propagation of plants that are resistant to the disease or pest.

Forest health management tools and practices can enhance or limit forest resilience and adaptation. Forest health management can be quite complicated, as the various tools are only effective for certain pests; for example, chemical methods are often ineffective for controlling borers and root diseases, while a different biological control may be needed for each pest. Because of the variety of tools, it is difficult to categorize the effects of the tools on resilience and adaptability. Nonetheless, forest health can be improved generally, if the management tools are selected and implemented to remove invasive species and expand the native genetic, species, and stand diversity of forests.

Many observers distinguish between endemic or native species of insects and diseases and exotic or foreign pests, and suggest that eradicating exotic and invasive species would generally improve the ecological sustainability of ecosystems. However, changing conditions can lead native pests into new, adjoining ecosystems; for example, warmer winters have allowed the mountain pine beetle to migrate into jack pine stands of northern Alberta, where it has not occurred previously. Should an endemic species that moves into a new ecosystem be considered an exotic species, or should this migration be considered natural? This is an important but difficult determination for those who support eliminating invasive species.

Timber Production and Management

Timber production and management, or silviculture, has been an historically important part of forest management. Silvicultural decisions determine the timber harvesting, thinning, and reforestation practices employed on forest land—industrial plantations as well as private woodlots and publicly owned forests. [99] These decisions affect the diversity of both existing and future forests by choosing which (if any) trees to cut, which trees to let grow, and in what patterns, thus determining the genetic, species, and stand diversity of forests. In addition, forest management certification is a means to assess and assure sustained forests over the long term.

Timber Harvesting

Timber harvesting is cutting and removing trees, usually to produce wood products (lumber, plywood, paper, etc.) Decisions about the cutting system to use, the rotation age (the age or size when trees is to be cut), and cutting unit size and dispersal affect forest diversity at all levels. Standard cutting systems typically yield even-aged stands, where all or most of the trees are in one or two age classes, and uneven- or all-aged stands, where the trees vary in age and size.[100] All-aged stands are generally more asynchronous and often have greater tree species diversity, but even-aged stands are natural for many temperate species, especially those established through stand replacement fires or other forest-level catastrophic disturbances.

The selection of which trees to cut is a critical aspect of the harvesting decision. Historically, the largest and straightest trees of particular species were preferred, because they produce the most and best wood products. However, leaving smaller trees, those with poorer form, and the less desirable species may result in a degraded forests, especially if these are the trees providing seed for future forests. Salvage and sanitation harvests, however, commonly focus on removing dead, dying, infected/infested, and at-risk trees, and thus can be a component of forest health management. (See above.) Leaving "legacy trees," with their superior growth characteristics and/or resistance to particular insects or diseases,[101] is particularly important, because they can provide seeds for regenerating adaptable forests, as well being healthy, resilient trees themselves.

Rotation age depends largely on the purpose of the forest: industrial plantations may maximize timber production, while conservation areas will likely emphasize tree sizes to maximize other ecosystem services. Lower rotation ages can speed forest adaptation by allowing species better adapted to climate change to regenerate (see below), while reducing forest synchrony and the period of time forests are vulnerable to pests and disease.[102] However, shorter rotation ages also eliminate some ecosystem values associated with very old trees (e.g., habitat for certain animal species and carbon sequestration) and reduces overall stand asynchrony by resulting in fewer age classes of tree stands.

Similarly, cutting unit size generally depends on the purpose of the forest. Larger sizes (40 acres or more for many even-aged forests) generally lower management costs and yield greater harvests in a single cutting operation. Small cutting units can increase biodiversity and stand asynchrony, and thus can contribute to resilient, adaptable forests. However, small cutting units can also cause problems for healthy forests, because they require more roads to remove the same amount of biomass. Roads are problematic for forest management, as road construction and inadequate road maintenance are primary sources of erosion and stream sedimentation. In addition, smaller cutting units and more roads contribute to forest fragmentation. (See "Fragmentation and Corridors," below.) All-aged management can intensify the road and fragmentation problems of small cutting units for healthy forests.

Thinning

Thinning is the practice of cutting and removing a portion of the trees in a stand. It is commonly done to enhance timber growth, to reduce wildfire threats by eliminating some fuels (especially fuel ladders); to halt or slow the spread of insect or disease infestations; to remove invasive species; to improve tree vigor; and to provide economic benefits. Because the biomass to be cut and removed is selected, thinning can also be used to enhance forest resilience and adaptability, by encouraging trees that are resistant to disturbances and that can recover more quickly when disturbances occur. At the broader scale, thinning can be used to enhance stand diversity, altering forest structure by removing trees to decrease stand density and uniformity, to increase forest asynchrony, and to expand the age, size, and species diversity of stands.[103] Thinning is not a panacea, however. Many object to the potential impacts of mechanical treatments—soil erosion and water quality degradation; removal of valuable wood rather than improving forest health; short-term increase in fire risk from the small-diameter biomass that remains; and more.[104] On the other hand, while the wood from thinning may have little value for traditional wood products, these activities can increase the supply of wood for biomass energy, offsetting consumption of fossil fuels.

Natural Forest Regeneration

Natural regeneration, generally following a timber harvest, relies on seeds from trees on lands surrounding the cutting units or uncut trees within the cutting units. It can help produce resilient and adaptable forests by providing genetic diversity for new stands, since many trees can provide seed sources for the regrown forest. Also, if the legacy trees and surrounding forest contain a mix of species, the regeneration is likely to reflect that species diversity. However, if historic and current harvests have removed all or most of the desirable (both in species and genetic characteristics) trees, natural regeneration might result in an unhealthy, genetically impoverished stand. Also, naturally regenerated trees sometimes takes several years to become established, which can allow invasive and non-tree species to dominate a site.

Artificial Forest Regeneration

Planting trees established in nurseries, called artificial regeneration, is a common alternative to natural regeneration. Forest management, especially for industrial plantations, has traditionally focused on retaining a few desired, previously established species. However, planting species and species mixes could promote resilient, adaptable forests. Suitability of species for planting could be related to resistance to pests, fire, drought, and other forest stressors. For example, aspen is a relatively short-lived species that recovers quickly from even the most intense fires; thus it could be planted among species slower to recover after fire in areas where fire is expected to increase.[105] Because the understanding of future conditions can be ambiguous or uncertain, reforesting with species that are successful under a wide range of conditions seems likely to enhance reforestation success.[106] Caution may be warranted, however, when reforesting with species and species mixes in anticipation of climate change. Some have suggested the need to plant trees beyond (further north and further upslope from) the situations where the species has been planted,[107] but such planting could cause unexpected and unintended consequences, due to the complex nature of ecosystems and species interactions. To the extent that such planting is outside traditional practices, experimentation and monitoring may be necessary to improve the probability of regeneration success without unintended harm to ecosystem services.[108]

Tree nurseries could also contribute to adaptation by providing genetically appropriate planting stock. Trees can be bred for tolerance to stressors. Genetic options are already available for some species, including for drought hardiness, delayed or advanced growth initiation, and pest resistance.[109] Genetic engineering (e.g., genetically modified organisms, or “GMOs”) might also be useful, but could be controversial.

Forest Management Certification

Certification of forest management as sustainable might also contribute to forests that are adaptable and resilient. Forest certification organizations emphasize forest sustainability, including low-impact logging practices, and their certifications provide financial incentives (higher wood product prices and sometimes revenues from other ecosystem services) to landowners that encourage biodiversity and other practices that can assist adaptation. Certification is a process by which an independent organization examines forest management according to certain established principles of sustainable forestry to assure that the certified forests will continue to produce the desired ecosystem services (including wood products). There are several sustainable forest certification organizations, including the Forest Stewardship Council (FSC), the Sustainable Forestry Initiative (SFI), the Central Point of Expertise on Timber Procurement (CPET), and the Canadian Standards Association (CSA).[110]

Low-impact logging is a particular practice encouraged by certification that could contribute to sustainable, resilient forests, It is a collection of practices is intended to minimize forest disturbances from harvesting.[111] Although developed largely for tropical forests, where logging can cause extensive damage to the residual trees, low-impact logging practices are relevant to U.S. temperate forests for resilience and adaptation. Some practices, such as designing cutting units to minimize road needs and to protect residual stands, are relatively commonplace in the United States. However, other low-impact logging practices which can aid adaptation are relatively less common, such as leaving legacy-trees, reducing the water quality impacts of road construction and maintenance, and adjusting rotation intervals.

Fragmentation and Corridors

Fragmented habitats are habitat parcels separated by stretches of other habitat, such as forest patches separated by farmland or housing developments.[112] Species in fragmented habitats have physical barriers to dispersal. Many plant and animal species cannot move readily across farmland, highways, or other breaks in their habitat. Habitat fragmentation adds to the difficulties when ecological habitat shifts out-pace natural species migration rates.[113] (See "Limits to Natural Adaptation," above.) Climate change might move suitable habitats faster than many species can disperse; fragmentation of habitats makes it more difficult for species to disperse to new suitable habitats.[114]

Habitat corridors can eliminate some dispersal barriers.[115] Corridors are landscapes that contain continuous habitat with relatively few physical or biotic barriers to species migration. For most of the continental United States, broad habitat corridors are now fragmented by human land-use developments. Fragmented habitats along migration pathways can sometimes be connected through policies and practices to acquire lands and influence land management and to maintain connectivity where it is threatened. Some suggest designing landscapes to facilitate species spread.[116] Others propose managing landscapes at large scales to maximize connectivity.[117] Both may be difficult, because migration pathways may be more restricted in some places than others, depending on topography and other barriers to dispersal. It can be also difficult because such efforts encompass encouraging multiple landowners to act in coordinated, cohesive ways and may require innovative approaches to encourage species dispersal over a multitude of land uses. It may also require a politically challenging array of regulations and incentives to achieve large-scale connectivity and corridors. Nonetheless, habitat corridors are considered to be one of the least risky adaptation management strategies.[118]

Existing species have migrated or adapted to accommodate climate change in the past. However, given the increasing habitat fragmentation of the past two centuries, it seems likely that restricted natural dispersal of species will result in the extinction of species that might have otherwise survived.[119] Although forests may eventually be productive and valuable after periods of transition, some animal species will likely be lost if corridors do not connect them with suitable habitats.

Managed Relocation

Managed relocation is another management strategy for assisting forest adaptation. Managed relocation (also known as assisted migration or translocation) is the intentional movement of trees and related species to locations where the probability of future success is predicted to be higher.[120] It is akin to artificial forest regeneration, discussed above, but encompasses the regeneration of whole ecological systems, based on the premise that natural migration may not be able to keep pace with climate change.

Many experts believe managed relocation is a high-risk strategy for adaptation.[121] Due to the complexity of ecosystem community interactions and relationships, managed relocation efforts may require multi-species relocation. Some species depend on others for survival, with relationships ranging from mutualism or symbiosis (where both species benefit) to parasitism (where one species essentially lives off the other). Such community interactions would need to be understood in the ecosystems where species are relocated, to better understand the impacts and feasibility of relocation.[122] Some advocates recommend experimentation with species introductions over a range of habitats, rather than solely within historical habitat, followed by monitoring to assess success.[123]

Effective managed relocation of ecosystems may require more accurate projections of habitat shifts than currently exist. Paleo-ecological data (from pollen records, tree rings, charcoal deposits, animal fossils, and more) and records of historical species ranges may provide insight into forest responses to climate change and natural species migration over longer periods.[124] For example, northern populations of many temperate forest types may be pre-adapted to climate change from previous exposure to population expansion after the last ice age, and might therefore be a logical option for relocation north of current habitats.[125]

Some argue that the species introduced to new areas through managed relocation would be invasive species and the uncertainties about the possible success and impacts of relocation should restrain relocation efforts.[126] Managed relocation clearly can disrupt the communities where relocation is being implemented.[127] Although, ecologists have made a concerted effort to study invasive species, reliable predictions about invasive species dominance remain elusive. It can take decades for invasive species to dominate the ecosystems into which they have been introduced.[128] This suggests efforts to monitor relocated species may be difficult, costly, and long-term. Future research could emphasize species' thresholds to trigger relocation, species prioritization, and best practices for relocation to minimize damage to host ecosystems and the ecosystem services they provide. However, for landowners with extensive holdings, such as some major forest products companies, managed relocation of desirable timber species might be a practical, if expensive, approach to assuring timber supplies and other key ecosystems services in the long term.

End Notes

[1] J. McLachlan et al., "A Framework for Debate of Assisted Migration in an Era of Climate Change," *Conservation Biology*, vol. 21, no. 2 (2007), pp. 297-302.

[2] L. Joyce et al., "Chapter 17: Potential Consequences of Climate Variability and Change for the Forests of the United States," in *Climate Change Impacts on the United States*, ed. National Assessment Synthesis Team (Cambridge: Cambridge Univ. Press, 2001), pp. 489-524.

[3] This report does not address managing forests to mitigate climate change (i.e., managing forests to increase carbon sequestration), although many of the forestry practices can enhance carbon sequestration and thus reduce the impacts of carbon emissions on climate. For more information, see CRS Report RL31432, *Carbon Sequestration in Forests*; CRS Report RL34560, *Forest Carbon Markets: Potential and Drawbacks*; and CRS Report R41144, *Deforestation and Climate Change*.

[4] For a more thorough summary of federal efforts by the Forest Service and other federal departments and agencies, see T. Cruce and H. Holsinger, *Climate Change Adaptation: What Federal Agencies Are Doing*, Pew Center on Global Climate Change, Washington, DC, November 2010, http://www.pewclimate.org/docUploads/Federal-Government-Leadership-Adaptation-Nov-2010.pdf.

[5] See http://www.fs.fed.us/climatechange/pdf/roadmap.pdf.

[6] C. Allen et al., "A Global Overview of Drought and Heat-Induced Tree Mortality Reveals Emerging Climate Change Risks for Forests," *Forest Ecology and Management*, vol. 259 (2010), pp. 660-684.

[7] P. Gonzalez et al., "Forest Carbon Densities and Uncertainties from Lidar, QuickBird, and Field Measurements in California," *Remote Sensing of Environment*, vol. 114, (2010), pp. 1561-1575.

[8] For a discussion of the science of climate change, see CRS Report RL34266, *Climate Change: Science Highlights*.

[9] D. Spittlehouse and R. Stewart, "Adaptation to Climate Change in Forest Management," *BC Journal of Ecosystems and Management*, vol. 4, no. 1 (2003).

[10] Joyce et al., "Consequences of Climate Change for U.S. Forests."

[11] G. Tang and B. Beckage, "Projecting the Distribution of Forests in the New England Response to Climate Change," *Diversity and Distributions*, vol. 16, (2010), pp. 144-158.

[12] B. Beckage et al., "A Rapid Upward Shift of a Forest Ecotone During 40 Years of Warming in the Green Mountains of Vermont," *PNAS*, vol. 105, no. 11 (March 8, 2008), pp. 4197-4202.

[13] A. Evans and R. Perschel, "A Review of Forestry Mitigation and Adaptation Strategies in the Northeast U.S.," *Climatic Change*, vol. 96, no. 1-2 (2009), pp. 167-183.

[14] Tang and Beckage, "Distribution of Forests in the New England Response to Climate Change."

[15] Gonzalez et al., "Ecosystem Vulnerability to Climate Change."

[16] S. Crimmins et al., "Changes in Climatic Water Balance Drive Downhill Shifts in Plant Species' Optimum Elevations," *Science*, vol. 331 (January 21, 2011), pp. 324-327.

[17] G. Meehl et al., "Chapter 10: Global Climate Projections," in *Climate Change 2007: The Physical Science Basis. Contribution of Working Group I to the Fourth Assessment Report of the Intergovernmental Panel on Climate Change* (New York: Cambridge Univ. Press, 2007), p. 750.

[18] K. Redmond, "Climate Variability and Change as a Backdrop for Western Resource Management," in *Bringing Climate Change Into Natural Resource Management: Proceedings of a Workshop*, Gen. Tech. Rept. PNW-GTR-706 (Portland, OR: USDA Forest Service, March 2007), pp. 5-40.

[19] National Assessment Synthesis Team, *Climate Change Impacts on the United States: Potential Consequences of Climate Variability and Change—Overview: Northeast* (Washington: U.S. Global Climate Change Research Program, 2000), pp. 40-45.

[20] Joyce et al., "Consequences of Climate Change for U.S. Forests," p. 501.

[21] Allen et al., "Climate Change Risk for Forests."

[22] Spittlehouse and Stewart, "Forest Adaptation to Climate Change."

[23] D. McKenzie et al., "Climatic Change, Wildfire, and Conservation," *Conservation Biology*, vol. 18, no. 4 (2004), pp. 890-902.

[24] Allen et al., "Climate Change Risk for Forests."

[25] USDA Forest Service, *Major Forest Insect and Disease Conditions in the United States 2008 Update*, FS-933, Washington, DC (Sept. 2009).

[26] J. Logan et al., "Assessing the Impacts of Global Warming on Forest Pest Dynamics," *Frontiers in Ecology and the Environment*, vol. 1, no. 3 (2003), pp. 130-137.

[27] Logan et al., "Global Warming and Forest Pest Dynamics."

[28] See CRS Report R40203, *Mountain Pine Beetles and Forest Destruction: Effects, Responses, and Relationship to Climate Change*.

[29] D. Leatherman et al., *Mountain Pine Beetle*, Colorado State Univ. Extension, no. 5.528, at http://www.colostate.edu/ pubs/insect/05528.html.

[30] D. Leatherman et al., *Mountain Pine Beetle*.

[31] McKenzie et al., "Climatic Change, Wildfire, and Conservation."

[32] J. Littell et al., "Climate and Wildfire Area Burned in Western U.S. Ecoprovinces, 1916-2003", *Ecological Applications*, vol. 19, no. 4 (2009), pp. 1003-1021; McKenzie et al., "Climatic Change, Wildfire, and Conservation."

[33] Joyce et al., "Consequences of Climate Change for U.S. Forests."

[34] A. Westerling et al., "Warming and Earlier Spring Increase Western U.S. Forest Wildfire Activity," *Science*, vol. 313 (August 18, 2006), pp. 940-943.

[35] P. Backlund et al., *The Effects of Climate Change on Agriculture, Land Resources, Water Resources, and Biodiversity in the United States*, Report by the U.S. Climate Change Science Program and the Subcommittee on Global Change Research, Synthesis and Assessment Product 4.3, Washington, DC, May 2008, p. 5, http://www.globalchange.gov/publications/reports/scientific-assessments/saps/sap4-3.

[36] M. Kauffman et al., *The Status of Our Scientific Understanding of Lodgepole Pine and Mountain Pine Beetles—A Focus on Forest Ecology and Fire Behavior*, GFI Tech. Rept. 2008-2 (Arlington, VA: The Nature Conservancy, 2008).

[37] J. Aber et al., "Forest Processes and Global Environmental Change: Predicting the Effect of Individual and Multiple Stressors," *BioScience*, vol. 51, no. 9 (September 2001), pp. 735-751.

[38] Joyce et al., "Consequences of Climate Change for U.S. Forests."

[39] Evans and Perschel, "Forestry Mitigation and Adaptation Strategies in the Northeast."

[40] Joyce et al., "Consequences of Climate Change for U.S. Forests."

[41] Joyce et al., "Consequences of Climate Change for U.S. Forests."

[42] S. McMahon et al., "Evidence for a Recent Increase in Forest Growth," *PNAS*, vol. 107, no. 8 (February 23, 2010), pp. 3611-3615.

[43] C. Millar et al., "Climate Change and Forests of the Future: Managing in the Face of Uncertainty," *Ecological Applications*, vol. 17, no. 8 (2007), pp. 2145-2151.

[44] S. Running and L. Mills, *Terrestrial Ecosystem Adaptation*, Resources for the Future, Washington, DC, June 2009, p. 11, http://www.rff.org/RFF/Documents/RFF-Rpt-Adaptation-RunningMills.pdf.

[45] Backlund et al., *The Effects of Climate Change on Agriculture, Land Resources, Water Resources, and Biodiversity*, Synthesis and Assessment Product 4.3, p. 9.

[46] Millar et al., "Climate Change and Forests of the Future."

[47] G. Blate et al., "Adapting to Climate Change in United States National Forests," *Unasylva 231/232*, vol. 60 (2009), pp. 57-62.

[48] B. Kinloch and G. Dupper, "Genetic Specificity in the White Pine—Blister Rust Pathosystem," *Phytopathology* vol. 92, no. 3 (2002), pp. 278-280.

[49] See the American Chestnut Foundation webpage, http://www.acf.org/history.php.

[50] See M. Altieri, *Biodiversity and Pest Management in Agroecosystems* (Binghamton, NY: Haworth Press, 1994); and D. Pimentel and A. Wilson, "Non-Indigenous Species: Crops and Livestock," in *Encyclopedia of Pest Management*, ed. D. Pimentel, vol. 2 (Cleveland, OH: CRC Press, 2007), pp. 400-403.

[51] U.S. Dept. of Agriculture, Forest Service, *The Process Predicament: How Statutory, Regulatory, and Administrative Factors Affect National Forest Management*, Washington, June 2002, http://www.fs.fed.us/projects/documents/ Process-Predicament.pdf.

[52] See CRS Report RL31065, *Forestry Assistance Programs*.

[53] This is not to suggest that the federal government has no influence over private forest management. Many laws, such as the Clean Water Act and the Endangered Species Act, affect the opportunities and responsibilities of private forest landowners, but there is no federal regulation of nonfederal forestry practices, per se.

[54] See CRS Report RL33917, *Forestry in the 2008 Farm Bill*.

[55] See CRS Report R40911, *Comparison of Climate Change Adaptation Provisions in S. 1733 and H.R. 2454*.

[56] Millar et al., "Climate Change and Forests of the Future."

[57] B. St. Clair and G. Howe, "Genetic Options for Adapting Forests to Climate Change," *Western Forester*, vol. 54, no. 1 (Jan/Feb 2009), pp. 9-11.

[58] B. Griffith et al., "Climate Change Adaptation for the US National Wildlife Refuge System," *Environmental Management*, vol. 44, (2009), pp. 1043-1052.

[59] McLachlan et al., "A Framework for Debate of Assisted Migration in an Era of Climate Change."

[60] R. Keane et al., "The Use of Historical Range and Variability (HRV) in Landscape Management," *Forest Ecology and Management*, vol. 258, (2009), pp. 1025-1037.

[61] Millar et al., "Climate Change and Forests of the Future."

[62] J. Lawler, "Resource Management in a Changing and Uncertain Climate," *Frontiers in Ecology and the Environment*, vol. 8, no. 1 (2010), pp. 35-43.

[63] Griffith et al., "Adaptation for Wildlife Refuges;" McLachlan et al., "Assisted Migration in an Era of Climate Change."

[64] Blate et al., "Adapting to Climate Change in United States National Forests."

[65] Allen et al., "Climate Change Risk for Forests"; McMahon et al., "Evidence for a Recent Increase in Forest Growth."

[66] Lawler, "Resource Management in a Changing and Uncertain Climate"; P. Landres et al., "Overview of the Use of Natural Variability Concepts in Managing Ecological Systems," *Ecological Applications*, vol. 9, no. 4 (1999), pp. 1179-1188.

[67] T. DeLuca et al., "The Unknown Trajectory of Forest Restoration: A Call for Ecosystem Monitoring," *Journal of Forestry*, vol. 108 (September 2010), pp. 288-295.

[68] Millar et al., "Climate Change and Forests of the Future."

[69] Backlund et al., *The Effects of Climate Change on Agriculture, Land Resources, Water Resources, and Biodiversity*, Synthesis and Assessment Product 4.3, p. 4.

[70] See FIA website, http://fia.fs.fed.us/.

[71] McLachlan et al., "Assisted Migration in an Era of Climate Change."

[72] D. Durrant et al, *European Forests and the Environment: The Work and Achievements of EU/ICP Forests*, http://www.icp-forests.org/pdf/folder.pdf.

[73] See E. Tuchmann, *The Northwest Forest Plan: A Report to the President and Congress* (Portland, OR: USDA Forest Service, 1996).

[74] L. Joyce et al., "Chapter 3—National Forests," in *Preliminary Review of Adaptation Options for Climate-Sensitive Ecosystems and Resources*, Final Report, Synthesis and Assessment Product 4.4, U.S. Climate Change Science Program, Washington, DC, June 2008, http://www.climatescience.gov/Library/sap/sap4-4/final-report/.

[75] B. Bormann et al., "Adaptive Management of Forest Ecosystems: Did Some Rubber Hit the Road?" *BioScience*, vol. 57, no. 2 (2007), pp. 186-191.

[76] McLachlan et al., "Assisted Migration in an Era of Climate Change."

[77] Millar et al., "Climate Change and Forests of the Future."

[78] See CRS Report R41671, *Adaptive Management for Ecosystem Restoration: Analysis and Issues for Congress*.

[79] J. Anderson et al., "Watershed Restoration—Adaptive Decision Making in the Face of Uncertainty," in *Strategies for Restoring River Ecosystems: Sources of Variability and Uncertainty in Natural and Managed Systems*, ed. R. Wissmar et al. (Bethesda, MD: American Fisheries Society, 2003), pp. 205-206.

[80] Joyce et al., "Consequences of Climate Change for U.S. Forests."

[81] Gonzalez et al., "Global Patterns in the Vulnerability of Ecosystems Due to Climate Change."

[82] See Cruce and Holsinger, *Climate Change Adaptation: What Federal Agencies Are Doing*.

[83] See http://www.fs.fed.us/climatechange/pdf/roadmap.pdf.

[84] See http://www.usgs.gov/global_change/doi_taskforce.asp.

[85] See http://www.doi.gov/whatwedo/climate/index.cfm.

[86] M. Bekker and A. Taylor, "Fire Disturbance, Forest Structure, and Stand Dynamics in Montane Forests of the Southwestern Cascades, Thousand Lakes Wilderness, California, USA," *Ecoscience*, vol. 17, no. 1 (2010), pp. 59-72; McKenzie et al., "Climatic Change, Wildfire, and Conservation."

[87] Some observers have noted that protecting structures from wildfire in the wildland-urban interface is largely unrelated to wildfire management. For more information on protecting homes from wildfire, see CRS Report RL34517, *Wildfire Damages to Homes and*

Resources: Understanding Causes and Reducing Losses, and CRS Report RS21880, *Wildfire Protection in the Wildland-Urban Interface*.

[88] Bekker and Taylor, "Fire Disturbance, Forest Structure, and Stand Dynamics in Montane Forests."

[89] Christopher Lancette, "Poll Shows Changing Attitudes Toward Wildfire," The Wilderness Society, Oct. 22, 2008, http://wilderness.org/content/poll-shows-changing-attitudes-toward-wildfire.

[90] National Fire and Aviation Executive Board, Memorandum to: fire Management, Subject: Clarification of Appropriate Management Response, June 20, 2007.

[91] FUSEE (Firefighters United for Safety, Ethics and Ecology), *Issue Paper: Implementing Appropriate Management Response (AMR)*, Eugene, OR, http://fusee.org/docs/AMR/AMR%20Issue%20Paper.pdf.

[92] N. Berg and D. Azuma, "Bare Soil and Rill Formation Following Wildfires, Fuel Reduction Treatments, and Pine Plantations in the Southern Sierra Nevada, California, USA," *International Journal of Wildland fire*, v. 19 (2010), pp. 478-489; and Blate et al., "Adapting to Climate Change in United States National Forests."

[93] See CRS Report R40811, *Wildfire Fuels and Fuel Reduction*.

[94] Spittlehouse and Stewart, "Forest Adaptation to Climate Change."

[95] C. Wiedinmyer and M. Hurteau, "Prescribed Fire as a Means of Reducing Forest Carbon Emission in the Western United States," *Environmental Science and Technology*, vol. 44, no. 6 (2010), pp. 1926-1932.

[96] U.S. Dept. of Agriculture, Forest Service, Forest Health Protection, "Managing Native Insects & Diseases," http://www.fs.fed.us/foresthealth/management/fhm-natives.shtml.

[97] U.S. Dept. of Agriculture, Forest Service, Forest Health Protection, Pesticide Management & Coordination, "Health & Safety," http://www.fs.fed.us/foresthealth/pesticide/health.shtml.

[98] U.S. Dept. of Agriculture, Forest Service, Forest Health Protection, "Biological Control," http://www.fs.fed.us/ foresthealth/biologicalcontrol/index.shtml.

[99] See D. Smith et al., *The Practice of Silviculture: Applied Forest Ecology*, 9th ed. (New York: John Wiley & Sons, Inc., 1997).

[100] See CRS Report 98-917, *Clearcutting in the National Forests: Background and Overview*.

[101] J. Agee and C. Skinner, "Basic Principles of Forest Fuel Reduction Treatments," *Forest Ecology and Management*, vol. 211 (2005), pp. 83-96.

[102] Spittlehouse and Stewart, "Forest Adaptation to Climate Change."

[103] Spittlehouse and Stewart, "Forest Adaptation to Climate Change."

[104] A full discussion of the desirable and undesirable effects of thinning, as well as of logging generally, is beyond the scope of this report. For more information, see pages 9-17 of CRS Report RL31432, *Carbon Sequestration in Forests*.

[105] Spittlehouse and Stewart, "Forest Adaptation to Climate Change."

[106] Spittlehouse and Stewart, "Forest Adaptation to Climate Change."

[107] Millar et al., "Climate Change and Forests of the Future."

[108] Millar et al., "Climate Change and Forests of the Future."

[109] St.Clair and Howe, "Genetic Options for Adapting Forests."

[110] See WWF/World Bank Global Forest Alliance, *Forest Certification Assessment Guide (FCAG): A Framework for Assessing Credible Forest Certification Systems/Schemes*, Washington, DC, June 2006, http://assets.panda.org/ downloads/fcagfinal.pdf.

[111] D. Dykstra, *Reduced Impact Logging: Concepts and Issues*, UN Food and Agriculture Organization [FAO] Corporate Document Repository, http://www.fao.org/docrep/005/ac805e/ac805e04.htm.

[112] See K. Riitters, "Fragmentation," in *Forest Resources of the United states, 2007*, ed. W.B. Smith et al. (Washington, DC: USDA Forest Service, 2009), pp. 22-25.

[113] R. Noss, "Beyond Kyoto: Forest Management in a Time of Rapid Climate Change," *Conservation Biology*, vol. 15, no. 3 (2002), pp. 578-590.

[114] O. Honnay et al, "Possible Effects of Habitat Fragmentation and Climate Change on the Range of Forest Plant Species," *Ecology Letters*, vol. 5, no. 4 (July 2002), pp. 525-530.

[115] Griffith, et al., "Adaptation for Wildlife Refuges."

[116] McLachlan et al., "Assisted Migration in an Era of Climate Change."

[117] Millar et al., "Climate Change and Forests of the Future."

[118] Lawler, "Resource Management in a Changing and Uncertain Climate;" and Blate et al., "Adapting to Climate Change in United States National Forests."

[119] McLachlan et al., "Assisted Migration in an Era of Climate Change."

[120] D. Richardson et al., "Multidimensional Evaluation of Managed Relocation," *PNAS*, vol. 106, no. 24 (2009), pp. 9721-9724.

[121] Richardson et al., "Multidimensional Evaluation of Managed Relocation."

[122] McLachlan et al., "Assisted Migration in an Era of Climate Change;" Spittlehouse and Stewart, "Forest Adaptation to Climate Change."

[123] Millar et al., "Climate Change and Forests of the Future."

[124] J. Thompson et al, "Is There Potential for the Historical Range of Variability to Guide Conservation Given the Social Range of Variability?" *Ecology and Society*, vol. 14, no. 1 (2008), online article 18.

[125] McLachlan et al., "Assisted Migration in an Era of Climate Change."

[126] Lawler, "Resource Management in a Changing and Uncertain Climate."

[127] McLachlan et al., "Assisted Migration in an Era of Climate Change."

[128] See CRS Report RL30123, *Invasive Non-Native Species: Background and Issues for Congress*, especially pages 7-10.

INDEX

#

20th century, 59, 70

A

access, 3, 9, 10, 29, 57
accountability, 10
accreditation, 9, 11, 14, 16, 17, 48
adaptability, viii, 51, 59, 70, 73, 75
adaptation, 52, 53, 54, 58, 60, 61, 62, 63, 65, 67, 68, 69, 73, 74, 76, 77, 78
aesthetic, 25, 41, 43
age, 29, 31, 42, 57, 74, 75, 79
agencies, 13, 14, 24, 26, 41, 68, 69, 71, 79
Alaska, 48, 65
ASI, 12, 16
assessment, 17, 42, 43, 44, 64
assets, 67, 83
atmosphere, 55, 58
audit, 4, 11, 12, 17, 18, 19, 22, 45
audits, 6, 9, 16, 20
authority, 2, 27
awareness, 64

B

balance sheet, 64
banks, 40
barriers, 77
base, 14, 31, 32, 52, 58
beetles, 57
benchmarks, 6
benefits, 3, 5, 10, 22, 26, 40, 41, 60, 65, 71, 75
biodiversity, 22, 23, 30, 33, 34, 35, 36, 58, 59, 64, 70, 74, 76
biological control, 32, 72, 73
biomass, 22, 31, 53, 57, 58, 61, 71, 75
biotic, 77
board members, 14
boreal forest, 5, 61, 65
building code, 9
Bureau of Land Management, 64, 68
burn, 70, 71, 72
businesses, 32

C

carbon, viii, 5, 25, 40, 42, 51, 52, 53, 58, 61, 68, 74, 79
carbon emissions, 79
catalyst, 19
catastrophes, 29
category a, 22, 25, 26, 40
certificate, 12, 13, 14, 17
certification, vii, 1, 2, 3, 4, 5, 6, 7, 9, 10, 11, 12, 13, 14, 16, 17, 18, 19, 20, 21, 25, 26, 27, 41, 42, 45, 48, 65, 73, 76, 77
challenges, 18

chemical, 22, 32, 72, 73
classes, 57, 63, 74
classification, 23
Clean Water Act, 81
climate, viii, 25, 42, 51, 52, 53, 54, 56, 57, 58, 60, 61, 62, 63, 65, 66, 67, 68, 69, 72, 74, 76, 78, 79, 80, 82
climate change, viii, 25, 42, 51, 52, 54, 56, 58, 60, 61, 62, 65, 66, 67, 68, 72, 74, 76, 78, 79, 80
climates, 61
climatic factors, 55
Clinton Administration, 64
closure, 29
CO2, 58, 67, 72
commercial, 19, 31, 63
communities, 33, 34, 35, 36, 37, 64, 79
community, 4, 5, 33, 34, 61, 78
compaction, 39
competition, 6, 9, 61
complement, 4
complexity, 4, 31, 78
compliance, vii, 1, 3, 6, 7, 12, 25, 26, 42
composition, 17, 59
conflict, 16, 40, 41
Congress, viii, 2, 3, 26, 27, 36, 45, 51, 52, 60, 82, 84
conifer, 55
connectivity, 23, 34, 62, 77
consensus, 18, 20, 46, 55
conservation, 5, 23, 29, 33, 34, 35, 36, 37, 40, 41, 42, 61, 68, 69, 74
conservation XE "conservation" programs, 69
conserving, 5, 23, 24
consolidation, 6
construction, 9, 29, 37, 38, 75, 77
consulting, 45
consumers, 8, 9
consumption, 75
containers, 45
control measures, 39
controversial, 76
conversations, 11
cooperation, 54, 68
cost, 19
covering, 42
crops, 59
crown fires, 71
CSA, 7, 77
CT, 47, 48
cultivation, 22
customer loyalty, 10
cycles, 5, 34, 63

D

damages, 70, 72
danger, 56
deforestation, 4, 9
degradation, 4, 75
degraded forest health, vii, 1, 2
Department of Agriculture, 2, 54, 67
Department of Defense, 3
Department of Energy, 3
Department of the Interior, 68
deposition, 58, 64
deposits, 79
depth, 65
destruction, 64
developing countries, 4
developing nations, 13
directives, 4
disappointment, 5
diseases, 53, 54, 57, 63, 64, 71, 72, 73, 74
distribution, 62
diversity, 5, 14, 16, 30, 33, 34, 35, 42, 53, 59, 60, 61, 62, 64, 67, 70, 72, 73, 74, 75
DOI, 68
dominance, 54, 79
drought, viii, 51, 54, 56, 57, 59, 71, 76

E

early warning, 64
Earth Summit, 4
ecological data, 43, 63, 78
ecological restoration, 61
ecological systems, 62, 65, 78
economic development, 45

economic losses, 31
economics, 32
ecosystem, viii, 5, 14, 22, 23, 24, 25, 29, 34, 51, 52, 53, 55, 58, 61, 62, 63, 67, 70, 73, 74, 76, 78, 79
education, 17, 44, 45
emerging markets, 31
employees, 42, 44
endangered, 23, 33
endangered species, 23, 33
energy, 53, 75
enforcement, 11
engineering, 76
England, 55, 80
environment, 44
environmental change, 64
environmental factors, viii, 51
environmental impact, 29, 44
environmental issues, 64
environmental management, 16
EPA, 38
epidemic, 57, 59, 60
equipment, 19, 28, 30
erosion, 22, 24, 30, 35, 38, 39, 40, 75
EU, 63, 82
Europe, 6
European Union, 63
evidence, 32, 35, 66
evolution, 9
expertise, 54, 69
exposure, 79
extinction, 58, 78

F

families, 42
Farm Bill, 81
farmland, 77
federal agency, 18
federal assistance, 53, 60, 69
federal government, 18, 54, 60, 67, 81
fiber, 16, 48
financial, viii, 32, 36, 51, 54, 60, 69, 76
financial incentives, 76
financial support, 36
fire suppression, 70
fires, viii, 51, 57, 70, 71, 72, 74, 76
fisheries, 40
fishing, 33
flexibility, 20, 30, 65
floods, 56
Food, Conservation, and Energy Act of 2008, 60
forest ecosystem, 23, 24, 25, 29, 52, 61, 62, 70
forest fire, 57
forest habitats, 52, 66
forest management, vii, viii, 1, 2, 3, 4, 5, 7, 10, 11, 14, 15, 16, 20, 21, 22, 23, 26, 27, 28, 33, 34, 35, 43, 44, 52, 53, 54, 60, 61, 63, 64, 65, 67, 73, 75, 76, 81
forest resources, 2, 30, 40, 41
Forest Stewardship Council (FSC), vii, 1, 5, 13, 14, 16, 17, 77
fossils, 79
fuel ladders, 75
fuel loads, 71
fuel management, 22, 31, 71
funding, viii, 51, 53, 60
funds, 60
fungi, 71
fungus, 59

G

genetic diversity, 59, 75
global warming, 57
goods and services, viii, 51, 52, 60, 61, 66, 67
governance, 7, 11, 14
governments, 2, 12, 13, 54, 67, 68
grades, 31
grassroots, 14
grazing, 2, 24, 25, 40, 70, 71
greenhouse, 54, 58
greenhouse gas, 54, 58
growth, 20, 22, 23, 24, 28, 29, 30, 31, 36, 37, 43, 56, 57, 58, 63, 64, 74, 75, 76
growth rate, 58, 63

guidance, 5, 15, 23, 29, 32, 38, 39, 44, 46, 60, 68
guidelines, 5, 24, 28, 38, 39, 45

H

habitat, viii, 1, 21, 22, 23, 31, 33, 34, 35, 39, 52, 53, 55, 61, 62, 63, 65, 66, 74, 77, 78
habitats, 2, 23, 31, 33, 34, 40, 41, 42, 53, 55, 57, 64, 66, 77, 78, 79
hardwood forest, 66
harvesting, 2, 10, 20, 21, 23, 28, 29, 30, 37, 39, 52, 65, 70, 72, 73, 74, 77
Hawaii, 48, 66
hazardous waste, 45
health, vii, 1, 2, 14, 19, 23, 24, 29, 32, 42, 53, 58, 60, 63, 65, 70, 72, 73, 74, 75, 83
health problems, 60
hemisphere, 13
heterogeneity, 71
highways, 77
hiring, 20
history, vii, 4, 58, 70, 81
homes, 54, 82
host, 52, 54, 59, 72, 79
housing, 77
human, 52, 72, 77
human development, 52
human health, 72

I

images, 48
impact assessment, 44, 45
impairments, 24
in transition, 67
income, 19
indigenous peoples, 14, 15, 45
individuals, 2, 13, 59
industries, 3
industry, 5, 6, 7, 9, 11, 12
infestations, 56, 71, 75
initiation, 76
insect epidemics, viii, 51
insects, 54, 57, 64, 71, 72, 73, 74
inspections, 17
integrity, 61
interest groups, 6
interface, 82
international standards, 16
inventory process, 34, 35, 36
investment, 19
investments, 4, 19
issues, vii, 4, 17, 20, 21, 24, 25, 39, 42, 43, 64

L

labeling, 5
lakes, 38
land tenure, 14, 15, 25
landings, 29
landscape, 19, 23, 29, 31, 33, 34, 35, 37, 64, 68
landscapes, 23, 77
laws, 15, 24, 25, 26, 31, 32, 35, 38, 42, 45, 60, 81
laws and regulations, 15, 24, 25, 26, 31, 42, 45
lead, 17, 55, 56, 58, 73
LEED, 9
legislation, viii, 52
Lepidoptera, 73
life cycle, 10
livestock, 2, 24, 40, 59
local conditions, 5, 15, 18
local government, 2, 54, 67, 68
logging, 10, 39, 44, 70, 71, 76, 77, 83
long-term shifts, viii, 51
low risk, 35
loyalty, 10

M

magnitude, 62
management, vii, viii, 1, 2, 3, 4, 5, 6, 7, 10, 11, 14, 16, 18, 19, 20, 21, 22, 23, 24, 25, 26, 27, 28, 30, 31, 32, 33, 34, 35, 36, 37, 38, 40, 41, 43, 44, 45, 46, 51, 52, 53, 54,

60, 61, 62, 63, 64, 65, 67, 68, 69, 70, 71, 72, 73, 74, 76, 77, 78, 81, 82, 83
management planning, viii, 1, 21, 25, 26, 41, 43, 45
mapping, 36
market access, 10
market incentives, 5
marketplace, 10
materials, 3, 28, 45
matter, 57
measurement, 63
membership, 10, 11, 12, 13, 14, 16
methodology, 34, 35
Mexico, 5
migration, 34, 52, 53, 61, 63, 70, 73, 77, 78, 79
Minneapolis, 5
missions, 54
models, 42, 43, 56, 57, 61
moisture, 56
mortality, 54, 56, 57, 63

N

National Forest System, vii, 2, 4, 18, 27
national forests, vii, 1, 2, 3, 4, 6, 7, 20, 21, 24, 25, 26, 27, 36, 37, 40, 41, 44, 46, 60
national policy, 5
native species, 30, 72, 73
natural disturbance, 22, 29, 57, 70
Natural Resources Conservation Service, 54, 69
negotiating, 7
New England, 55, 80
NGOs, 4, 5
niche market, 10
nitrogen, 58
nonfederal forestry assistance, viii, 52
North America, 6, 7, 11, 15, 48

O

objectivity, 19
OH, 81
oil, 5, 45
operations, viii, 1, 5, 7, 10, 11, 14, 16, 22, 24, 25, 28, 29, 30, 31, 33, 38, 40, 41, 44, 48
opportunities, 2, 41, 53, 61, 81
outreach, 45
oversight, 12, 14, 65
ownership, 15, 36, 37, 41, 43, 63
ozone, 58

P

Pacific, 47, 54, 64, 66, 68
parasites, 72
participants, 21
pathogens, 72
pathways, 77
permission, iv
pesticide, 83
pests, viii, 32, 51, 52, 57, 59, 63, 71, 72, 73, 74, 76
plant growth, 56, 58
plants, 41, 56, 59, 72
policy, 3, 4, 5, 15, 18, 22, 38, 64
pollen, 79
pollution, 64
population, 79
potential benefits, 3
precipitation, viii, 51, 52, 56, 58, 62, 64
predators, 72
preparation, 22, 28, 30, 39
prescribed burning, 52, 65, 71, 72
President, 82
prevention, 31, 32, 39
principles, 5, 13, 14, 15, 76
private sector, 3
probability, 76, 78
professionals, 14, 44
profit, 16
progress reports, 18, 44
project, 20, 56
propagation, 73
protected areas, 29, 31, 32, 33, 34
protection, viii, 1, 2, 23, 24, 31, 33, 34, 35, 37, 38, 39, 70
public awareness, 64

public safety, 31
public service, 38
public transparency requirements, vii, 1
publishing, 5
pulp, 9

R

rainfall, 56
rash, 10
rate of change, 53, 61
real estate, 10
reasoning, 10
recovery, 33, 53, 58
recreation, 2, 24, 40, 41, 52, 67
recreational, 41, 53, 61
redistribution, 57
regenerate, 57, 74
regeneration, 22, 23, 30, 53, 57, 65, 75, 76, 78
regulations, 3, 15, 22, 24, 25, 26, 31, 35, 42, 45, 78
regulatory requirements, 16, 29
rehabilitation, 22
requirements, vii, 1, 2, 3, 4, 7, 11, 14, 16, 17, 18, 19, 20, 22, 23, 24, 25, 26, 27, 29, 30, 32, 33, 34, 36, 41, 43, 44, 45, 46
researchers, 62
resilience, viii, 51, 53, 54, 58, 59, 60, 61, 62, 67, 70, 73, 75, 77
resistance, 59, 74, 76
resolution, 17, 18
resource management, 16, 21, 24, 43, 54, 68
resources, 2, 3, 5, 24, 25, 30, 38, 39, 40, 41, 42, 70, 71
response, 9, 17, 32, 45, 56, 58, 65, 68, 70, 71
restoration, 33, 37, 61
restrictions, 29
revenue, 54
rights, 14, 15, 25, 40, 41, 45
rings, 79
risk, 30, 31, 33, 35, 39, 71, 74, 75, 78
risks, vii, 1, 2, 4, 32, 58, 59
root, 73

S

safety, 14, 16, 20, 22, 26, 31, 42, 44
science, 65, 68, 80
scope, 83
sedimentation, 22, 24, 39, 75
seed, 22, 34, 35, 74, 75
seedlings, 35, 56
services, viii, 16, 36, 38, 40, 42, 51, 52, 53, 54, 55, 58, 60, 61, 66, 67, 70, 74, 76, 79
severe wildfires, viii, 51, 71
SFI, vii, 1, 2, 5, 6, 7, 8, 9, 10, 11, 12, 13, 14, 15, 16, 17, 18, 19, 20, 21, 22, 23, 24, 25, 26, 27, 28, 29, 30, 31, 32, 33, 34, 35, 36, 37, 38, 39, 40, 41, 42, 43, 44, 45, 46, 47, 48, 77
signals, 10
small businesses, 32
social interests, 7, 14
social regulation, 22
soil erosion, 75
solution, vii, 1, 3
specialists, 61
species, viii, 22, 23, 30, 31, 32, 33, 34, 35, 36, 39, 51, 52, 53, 55, 56, 57, 58, 59, 61, 62, 63, 64, 65, 66, 67, 70, 71, 72, 73, 74, 75, 76, 77, 78, 79
Spring, 81
stability, 58
stakeholder groups, 3, 7
stakeholder involvement, vii, 1, 43
stakeholders, vii, 1, 3, 4, 11
state, 2, 3, 13, 14, 18, 26, 33, 35, 42, 54, 60, 69
states, 5, 35, 54, 60, 66, 67, 68, 83
statistics, 44
statutes, 3
storage, 25, 40, 42, 45
storms, viii, 51, 56
stress, viii, 51, 56, 58, 63
stressors, 63, 76
structural changes, 6
structure, 3, 7, 11, 13, 23, 31, 37, 71, 75
style, 65
supervision, 44

suppliers, 9
supply chain, 7, 8, 11
suppression, 70, 71
surveillance, 17
survival, 57, 78
survival rate, 57
sustainability, 2, 4, 6, 7, 11, 20, 21, 58, 73, 76
sustainable forest management, vii, 1, 2, 3, 4, 5, 7, 10, 11, 14, 15, 16, 20, 21, 28
Switzerland, 6
symbiosis, 78
synchronization, 57

T

target, 25, 32, 72
Task Force, 68
technical assistance, viii, 51, 60
techniques, viii, 28, 30, 51
temperature, viii, 51, 52, 53, 55, 56, 58, 62, 66
tenure, 14, 15, 25, 40, 41
thinning, 19, 52, 65, 71, 73, 75, 83
threats, 60, 62, 75
timber harvests, vii, 1, 2, 20, 21, 28
timber production, 53, 64, 68, 74
time frame, 7, 23, 30, 43
trade, 4, 6, 60
traditional practices, 76
training, 16, 17, 19, 20, 44
transparency, vii, 1, 25
transportation, 32, 34, 39, 44
tropical forests, 5, 53, 59, 65, 77
tundra, 55

U

U.S. Department of Agriculture, 2
United Nations, 4
United States, vii, 1, 2, 4, 5, 6, 7, 8, 10, 13, 14, 15, 19, 22, 26, 33, 36, 47, 53, 55, 57, 61, 65, 66, 69, 77, 79, 80, 81, 82, 83, 84
updating, 26
USA, 82, 83
USDA, 3, 8, 54, 60, 63, 66, 67, 69, 80, 82, 83

V

valuation, 35
variations, 38, 52, 58, 59, 70
vegetation, 20, 24, 40, 71
volatility, 52, 54, 56, 57
vote, 13, 15
voting, 12, 13
vulnerability, 57

W

Washington, 6, 7, 26, 66, 68, 69, 79, 80, 81, 82, 83
waste, 22, 28, 45
water, viii, 1, 5, 24, 29, 30, 38, 39, 40, 41, 42, 51, 52, 53, 54, 58, 61, 63, 67, 75, 77
water quality, 24, 29, 38, 40, 42, 53, 58, 75, 77
water resources, 5, 30, 38, 39
wetlands, 24, 39
wilderness, 23, 36, 83
wildfire, vii, 1, 2, 31, 53, 56, 64, 70, 71, 72, 75, 82, 83
wildfire risks, vii, 1, 2
wildland, 71, 82
wildland-urban interface, 82
wildlife, viii, 1, 2, 23, 24, 31, 33, 34, 35, 41, 42, 53, 55, 61
Wisconsin, 26, 54, 68
wood, 2, 3, 5, 6, 9, 10, 11, 27, 31, 54, 63, 68, 74, 75, 76
wood products, 3, 5, 6, 9, 10, 54, 68, 74, 75, 76
workers, 14, 44
working groups, 12, 13, 15